A Primer of Lebesgue Integration

A Primer of Lebesgue Integration

H. S. Bear
Department of Mathematics
University of Hawaii at Manoa
Honolulu, Hawaii

ACADEMIC PRESS, INC.

San Diego New York Boston
London Sydney Tokyo Toronto

This text was typeset using the AMS TeX macros.

This book is printed on acid-free paper. ∞

Academic Press, Inc.
A Division of Harcourt Brace & Company
525 B Street, Suite 1900, San Diego, California 92101-4495

United Kingdom Edition published by
Academic Press Limited
24-28 Oval Road, London NW1 7DX

Library of Congress Cataloging-in-Publication Data

Bear, H. S. (Herbert Stanley)
A primer of Lebesgue integration / by Herbert S. Bear.
p. cm.
Includes index.
ISBN 0-12-083970-9
1. Lebesgue integral. I. Title.
QA312.B35 1995
515'.43--dc20 94-42043
CIP

PRINTED IN THE UNITED STATES OF AMERICA
95 96 97 98 99 00 BB 9 8 7 6 5 4 3 2 1

This text is dedicated to J. L. Kelley, who taught that in mathematics it is not enough to read the words—you've got to hear the music.

CONTENTS

PREFACE

This text provides an introduction to the Lebesgue integral for advanced undergraduates or beginning graduate students in mathematics. It is also designed to furnish a concise review of the fundamentals for more advanced students who may have forgotten one or two details from their real analysis course and find that more scholarly treatises tell them more than they want to know.

The Lebesgue integral has been around for almost a century, and the presentation of the subject has been slicked up considerably over the years. Most authors prefer to blast through the preliminaries and get quickly to the more interesting results. This very efficient approach puts a great burden on the reader; all the words are there, but none of the music. In this text we deliberately unslick the presentation and grub around in the fundamentals long enough for the reader to develop some intuition about the subject. For example, the Carathéodory definition of measurability is slick — even brilliant — but it is not intuitive. In contrast, we stress the importance of additivity for the measure function and so define a set $E \subset (0,1)$ to be measurable if it satisfies the absolutely minimal additivity condition: $m(E) + m(E') = 1$, where $E' = (0,1) - E$, and m is the outer measure in $(0,1)$. We then show in easy steps that measurability of E is equivalent to the Carathéodory criterion, $m(E \cap T) + m(E' \cap T) = m(T)$ for all T. In this way we remove the magic from the Carathéodory condition, but retain its utility. After the measure function is defined in $(0,1)$, it is extended to each interval $(n, n+1)$ in the obvious way and then to the whole line by countable additivity.

We define the integral via the familiar upper and lower Darboux sums of the calculus. The only new wrinkle is that now a measurable set is partitioned into a finite number of measurable sets rather than partitioning an interval into a finite number of subintervals. The use of upper and lower sums to define the integral is not conceptually different from the usual process of approximating a function by simple functions. However, the customary approach to the integral tends to create the impression that the Lebesgue integral differs from the Riemann integral primarily in the fact that the range of the function is partitioned rather than the domain. What is true is that a partition of the range induces an efficient partition of the domain. The real difference between the Riemann and Lebesgue integrals is that the Lebesgue integral uses a more sophisticated concept of length on the line.

We take pains to show that both the Riemann–Darboux integral and the Lebesgue integral are limits of Riemann sums, for that is the way scientists and engineers tend to think of the integral. This requires that we introduce the concept of a convergent net. Net convergence also allows us to make sense out of unordered sums and is in any case something every young mathematician should know.

After measure and integration have been developed on the line, we define plane outer measure in terms of coverings by rectangles. This early treatment of plane measure serves three purposes. First, it provides a second example of the definition of outer measure, and then measure starting with a natural geometric concept — here the area of a rectangle. Second, we show that the linear integral really is the area under the curve. Third, plane measure provides the natural concrete example of a product measure and is the prototype for the later development of general product measures.

The text is generously interlarded with problems. The problems are not intended as an intelligence test, but are calculated to be part of the exposition and to lure the reader away from a passive role. In many cases, the problems provide an essential step in the development. The step may be routine, but the reader is nevertheless encouraged thereby to pause and become actively involved in the process. There are also additional exercises at the end of each chapter, and the author earnestly hopes that these will add to the reader's education and enjoyment.

The author is pleased to acknowledge the help of Dick Bourgin, Bob Burckel, Ken Ross, all of whom read the manuscript with great care and suggested many improvements in style and content. In addition thanks are also due to Larry Wallen for much good advice.

1 THE RIEMANN--DARBOUX INTEGRAL

We start with the familiar Riemann–Darboux integral of the calculus, which for brevity we will call the Riemann integral. We consider a fixed bounded interval $[a, b]$, and consider only real functions f which are bounded on $[a, b]$.

A partition P of $[a, b]$ is a set $P = \{x_0, x_1, x_2, \ldots, x_n\}$ of points of $[a, b]$ with

$$a = x_0 < x_1 < x_2 < \cdots < x_n = b.$$

Let f be a fixed function on $[a, b]$ with

$$m \le f(x) \le M$$

for all $x \in [a, b]$. For each $i = 1, 2, \ldots, n-1$ let

$$\begin{aligned} m_i &= \inf\{f(x) : x_{i-1} \le x < x_i\}, \\ M_i &= \sup\{f(x) : x_{i-1} \le x < x_i\}. \end{aligned}$$

For $i = n$, let

$$\begin{aligned} m_n &= \inf\{f(x) : x_{n-1} \le x \le x_n\} \\ M_n &= \sup\{f(x) : x_{n-1} \le x \le x_n\}. \end{aligned}$$

In the usual calculus text treatment, the m_i and M_i are the infs and sups of values of $f(x)$ on the closed intervals $[x_{i-1}, x_i]$. Our later treatment of the Lebesgue integral will involve partitions into disjoint sets, so we make the intervals disjoint in this case too. This does not materially alter the definition.

The lower sum $L(f,P)$ and the upper sum $U(f,P)$ for the function f and the partition P are defined as follows:

$$L(f,P) = \sum_{i=1}^{n} m_i(x_i - x_{i-1}),$$

$$U(f,P) = \sum_{i=1}^{n} M_i(x_i - x_{i-1}).$$

Clearly $m \leq m_i \leq M_i \leq M$ for each i, so

$$m(b-a) \leq L(f,P) \leq U(f,P) \leq M(b-a).$$

For a positive function f on $[a,b]$ the lower sum represents the sum of the areas of disjoint rectangular regions which lie within the region

$$S = \{(x,y) : a \leq x \leq b, 0 \leq y \leq f(x)\}.$$

Similarly, the upper sum $U(f,P)$ is the area of a finite number of disjoint rectangular regions which cover the region S under the graph of f. The function f is said to be **integrable** whenever

$$\sup_P L(f,P) = \inf_P U(f,P). \tag{1}$$

The **integral** of f over $[a,b]$ is the common value in (1), and is denoted $\int_a^b f$. The area of S is defined to be this integral whenever f is integrable and non-negative.

The integral is also sometimes written $\int_a^b f(x)dx$, particularly when a change of variable is involved. The "x" in this expression is a dummy variable, and can be replaced by any variable except f. For example,

$$\int_a^b f(x)dx = \int_a^b f(t)dt = \int_a^b f(a)da = \int_a^b f(c)dc\,.$$

The last two versions are logically correct, but immoral, since they flout the traditional roles of a, b, c as constants, and the third integral discourteously uses the same letter for the limit and the dummy variable.

Now we make a few computations to derive the basic properties of the integral and show that the integral exists at least when f is continuous.

A partition Q is a refinement of the partition P provided each point of P is a point of Q. We will indicate this by writing $P \prec Q$ without reference to the numbering of the points in P or Q. Clearly Q is a refinement of P provided each of the subintervals of $[a,b]$ determined by Q is contained in one of the subintervals determined by P.

Proposition 1. *If* $P \prec Q$ *then* $L(f,P) \le L(f,Q)$ *and* $U(f,Q) \le U(f,P)$.

Proof. Suppose Q contains just one more point than P, and to be specific assume this additional point x^* lies between the points x_0 and x_1 of P. If

$$\begin{aligned} m_1' &= \inf\{f(x) : x_0 \le x < x^*\}, \\ m_1'' &= \inf\{f(x) : x^* \le x < x_1\}, \\ m_1 &= \inf\{f(x) : x_0 \le x < x_1\}, \end{aligned}$$

then $m_1' \ge m_1$ and $m_1'' \ge m_1$ so the sum of the first two terms in $L(f,Q)$ exceeds the first term of $L(f,P)$:

$$m_1'(x^* - x_0) + m_1''(x_1 - x^*) \ge m_1(x_1 - x_0).$$

The remaining terms of $L(f,Q)$ and $L(f,P)$ are the same, so $L(f,Q) \ge L(f,P)$. We can consider any refinement Q of P as obtained by adding one point at a time, with the lower sum increasing each time we add a point. The argument for the upper sums is similar. ▮

Proposition 2. *Every lower sum is less than every upper sum, as the geometry demands.*

Proof. If P and Q are any partitions, and $R = P \cup Q$ is the common refinement, then

$$L(f,P) \le L(f,R) \le U(f,R) \le U(f,Q). \quad ▮$$

Proposition 3. *f is integrable on $[a, b]$ if and only if for each $\varepsilon > 0$ there is a partition P of $[a, b]$ such that $U(f, P) - L(f, P) < \varepsilon$.*

Proof. This useful condition is equivalent to $\sup L(f, P) = \inf U(f, P)$ in view of Proposition 2. ▮

Proposition 4. *If f is integrable on $[a, b]$ and $[\alpha, \beta] \subset [a, b]$, then f is integrable on $[\alpha, \beta]$.*

Proof. Let $\varepsilon > 0$ and let P be a partition of $[a, b]$ such that $U(f, P) - L(f, P) < \varepsilon$. We can assume that α and β are points of P, since adding points increases $L(f, P)$ and decreases $U(f, P)$ and makes their difference smaller. If P_0 consists of the points of P which are in $[\alpha, \beta]$, then P_0 is a partition of $[\alpha, \beta]$. Note that

$$U(f, P) - L(f, P) = \sum_{i=1}^{m} (M_i - m_i)(x_i - x_{i-1}) \tag{2}$$

and $U(f, P_0) - L(f, P_0)$ is the sum of only those terms such that x_{i-1} and $x_i \in P_0$. Since we omit some non-negative terms from (2) to get $U(f, P_0) - L(f, P_0)$,

$$U(f, P_0) - L(f, P_0) \le U(f, P) - L(f, P) < \varepsilon . \quad ▮$$

Problem 1. If f is integrable on $[a, b]$, then $|f|$ is integrable on $[a, b]$ and $\left| \int_a^b f \right| \le \int_a^b |f|$. ◀

Problem 2. If $a < c < b$ then f is integrable on $[a, c]$ and on $[c, b]$ if and only if f is integrable on $[a, b]$. In this case

$$\int_a^c f + \int_c^b f = \int_a^b f . \quad ◀$$

Problem 3. If f is integrable on $[a, b]$ and $g = f$ except at a finite number of points, then g is integrable and $\int_a^b g = \int_a^b f$. ◀

Problem 4. If $a = x_0 < x_1 < \cdots < x_n = b$ and f is defined on $[a, b]$ with $f(x) = y_i$ for $x \in (x_{i-1}, x_i)$, then f is integrable and $\int_a^b f = \sum_{i=1}^{n} y_i(x_i - x_{i-1})$. (Note that it is immaterial how f is defined on the x_i by the preceding problem.) ◀

Problem 5. We say that g is a step function on $[a,b]$ if there is a partition $a = x_0 < x_1 < x_2 < \cdots < x_n = b$ such that g is constant on each (x_{i-1}, x_i). (By the preceding problem, step functions are integrable.) Show that if f is integrable on $[a,b]$ there are step functions g_n and h_n with $g_n(x) \le f(x) \le h_n(x)$ for all x and such that $g_n(x) \uparrow f(x)$ and $h_n(x) \downarrow f(x)$ for all but a countable number of points x, and $\lim_n \int_a^b g_n = \lim_n \int_a^b h_n = \int_a^b f$. ◀

Proposition 5. *If f is continuous on $[a,b]$ then f is integrable on $[a,b]$.*

Proof. If f is continuous on $[a,b]$, then f is uniformly continuous. Hence if $\varepsilon > 0$ there is $\delta > 0$ so that $|f(x) - f(x')| < \varepsilon$ whenever $|x - x'| < \delta$. If P is a partition with $x_i - x_{i-1} < \delta$ for all i, then $M_i - m_i \le \varepsilon$ for each i, so

$$\begin{aligned} U(f,P) - L(f,P) &= \sum (M_i - m_i)(x_i - x_{i-1}) \\ &\le \varepsilon \sum (x_i - x_{i-1}) \\ &= \varepsilon(b-a). \quad ∎ \end{aligned}$$

Problem 6. (i) If f is continuous on $[a,b]$ except at a (or b) and f is bounded on $[a,b]$, then f is integrable on $[a,b]$.

(ii) If f is bounded on $[a,b]$ and continuous except at a finite number of points, then f is integrable on $[a,b]$.

(iii) Suppose f is bounded on $[a,b]$ and discontinuous on a (possibly infinite) set E. Assume that for each $\varepsilon > 0$ there are disjoint intervals $(a_1, b_1), \ldots, (a_N, b_N)$ contained in $[a,b]$ such that $E \subset (a_1, b_1) \cup \cdots \cup (a_N, b_N)$ and $\sum_{i=1}^{N} (b_i - a_i) < \varepsilon$. Show that f is integrable. ◀

So far we have the integral $\int_a^b f$ defined only when $[a,b]$ is a bounded interval and f is bounded on $[a,b]$. Now we extend the definition to certain improper cases; *i.e.*, situations where the interval is unbounded, or f is unbounded on a bounded interval. Typical examples of such improper integrals are

$$\int_0^1 \frac{1}{\sqrt{x}}\, dx \quad \text{and} \quad \int_0^\infty \frac{1}{1+x^2}\, dx\,.$$

In both these examples the integrand is positive and the definition of the integral should give a reasonable value for the area under the curve. The definitions of the integrals above are:

$$\int_0^1 \frac{1}{\sqrt{x}}\,dx = \lim_{\varepsilon \to 0^+} \int_\varepsilon^1 \frac{1}{\sqrt{x}}\,dx\,,$$

$$\int_0^\infty \frac{1}{1+x^2}\,dx = \lim_{b \to \infty} \int_0^b \frac{1}{1+x^2}\,dx\,.$$

Both these limits are finite, so both functions are said to be (improperly) Riemann integrable on the given interval. The Lebesgue definition of the integral will give the same values.

In general, if f is integrable on $[a+\varepsilon, b]$ for all $\varepsilon > 0$, but f is not bounded on $[a, b]$ (*i.e.*, not bounded near a) we define

$$\int_a^b f = \lim_{\varepsilon \to 0^+} \int_{a+\varepsilon}^b f\,,$$

provided this limit exists. Similarly, if f is integrable on every interval $[a, b]$ for $b > a$ we define

$$\int_a^\infty f = \lim_{b \to \infty} \int_a^b f$$

when the limit exists. Similar definitions are made for $\int_a^b f$ if f is unbounded near b, and for $\int_{-\infty}^b f$ if f is integrable on $[a, b]$ for all $a < b$.

These definitions lend themselves to the calculations of elementary calculus, but do not coincide with the Lebesgue definition if f is not always positive or always negative. For example, if f is $(-1)^n/n$ on $[n, n+1)$, $n = 1, 2, \ldots$, then f is improperly Riemann integrable on $[1, \infty)$. We will see later that f is Lebesgue integrable if and only if $|f|$ is Lebesgue integrable. Hence the above function is not Lebesgue integrable since $\sum \frac{1}{n} = \infty$.

Problem 7. Show that $\int_0^\infty \frac{\sin x}{x}\,dx$ exists.

Problem 8. (i) Exhibit a g on $[0, \infty)$ such that $|g(x)| \equiv 1$ and $\int_0^\infty g$ exists.

(ii) Exhibit a function g on $[0, \infty)$ such that $|g(x)| \longrightarrow \infty$ as $x \longrightarrow \infty$ and $\int_0^\infty g$ exists.

(iii) Exhibit a function g on $[0, \infty)$ so that $|g(x)| \longrightarrow \infty$ and $\int_0^\infty g = 0$. Hint: Don't think about formulas for continuous functions, think about step functions.

Problem 9. Let $f(x) = x^2$ on $[0,1]$ and let $P_n = \{0, \frac{1}{n}, \frac{2}{n}, \dots, 1\}$. Write a formula for $L(f, P_n)$ and for $U(f, P_n)$. Show that $\int_0^1 x^2 dx = \frac{1}{3}$.

Problem 10. Write out the proof that $U(f,Q) \le U(f,P)$ whenever $P \prec Q$.

Problem 11. If f is integrable on $[a,b]$ and c is a number, then $\int_a^b cf = c\int_a^b f$. Show this for the case $c < 0$.

Problem 12. If f and g are integrable on $[a,b]$ and $f \le g$ on $[a,b]$, then $\int_a^b f \le \int_a^b g$.

Problem 13. Let $x \vee y = \max\{x,y\}$ and $x \wedge y = \min\{x,y\}$. For functions f and g, $(f \vee g)(x) = f(x) \vee g(x)$, $(f \wedge g)(x) = f(x) \wedge g(x)$. Show that if f and g are integrable on $[a,b]$ then $f \vee g$ and $f \wedge g$ are integrable, and $\int_a^b f \vee g \ge \int_a^b f \vee \int_a^b g$, $\int_a^b f \wedge g \le \int_a^b f \wedge \int_a^b g$.

2 THE RIEMANN INTEGRAL AS A LIMIT OF SUMS

The engineers and scientists who work with integrals think of the integral as a limit of sums:

$$\int_a^b f(x)dx = \lim \sum_{i=1}^{n} f(x_i)\Delta x_i \tag{1}$$

where $a = x_0 < x_1 < x_2 < \cdots < x_n = b$ is a partition of $[a, b]$ and $\Delta x_i = x_i - x_{i-1}$. Historically the integral sign is a flattened S for sum, and the "dx" suggests the typical (small) quantity Δx_i. The limit is the number that is approached as partitions are made finer and finer so that $\max \Delta x_i$ tends to zero.

Notice that the limit in (1) is a new animal. The sums we want the limit of depend on *partitions*. That is, we want the limit of a function of partitions, rather than a limit of a function of x or n. Assume for the moment that the limit in (1) makes sense for each of two functions f and g defined on $[a, b]$. Then

$$\begin{aligned}\int_a^b (f+g) &= \lim \sum (f(x_i) + g(x_i))\Delta x_i \\ &= \lim \left(\sum f(x_i)\Delta x_i + \sum g(x_i)\Delta x_i\right) \\ &= \lim \sum f(x_i)\Delta x_i + \lim \sum g(x_i)\Delta x_i \\ &= \int_a^b f + \int_a^b g.\end{aligned}$$

However we define the limit above, the limit of a sum is certainly the sum of the limits (*i.e.*, the third equality), for that is a basic property of addition: if a is close to A and b is close to B, then $a + b$ is close to $A + B$.

In this chapter we define a kind of limit, generalizing the idea of the limit of a sequence, which makes precise the ideas above. The objects we use to generalize sequences are called **nets**, and convergence of nets is sufficiently general to describe any kind of limit which occurs in analysis or topology.

A **directed set** is a non-empty set D equipped with a partial ordering $\prec$ satisfying the following conditions: (i) $\alpha \prec \alpha$ for all $\alpha \in D$,

(ii) if $\alpha \prec \beta$ and $\beta \prec \gamma$, then $\alpha \prec \gamma$,

(iii) for any elements α and β in D there is $\gamma \in D$ so $\alpha \prec \gamma$ and $\beta \prec \gamma$.

We will write $\beta \succ \alpha$ to mean the same as $\alpha \prec \beta$, and say that β is farther out than α when this relation holds.

A **net** is a function defined on a directed set.

A sequence is a type of net, with D being the set of natural numbers directed as usual: $n \prec m$ means $n \leq m$. We will adopt the sequence notation for nets and write $\{x_\alpha\}$ for the net consisting of the real valued function x defined on some directed set D of elements α. We will use some index other than n - for instance α - to emphasize that D need not be the set $\mathbb{N}$ of natural numbers.

The net $\{x_\alpha\}$ converges to ℓ, denoted $x_\alpha \longrightarrow \ell$, or $\lim_\alpha x_\alpha = \ell$, provided that for every $\varepsilon > 0$ there is $\alpha_0 \in D$ such that $|x_\alpha - \ell| < \varepsilon$ whenever $\alpha \succ \alpha_0$. Of course the limits of nets are unique if they exist, justifying the notation $\lim_\alpha x_\alpha = \ell$. The uniqueness is a consequence of the fact that there is some γ beyond any given α and β. Hence if ℓ_1 and ℓ_2 were different limits, with $\ell_2 - \ell_1 = \varepsilon > 0$, and $|x_\alpha - \ell| < \varepsilon/2$ when $\alpha \succ \alpha_0$ and $|x_\beta - \ell_2| < \varepsilon/2$ when $\beta \succ \beta_0$, then if $\gamma \succ \alpha_0$ and $\gamma \succ \beta_0$, the two contradictory conditions

$$|x_\gamma - \ell_1| < \varepsilon/2, \quad |x_\gamma - \ell_2| < \varepsilon/2$$

would both hold.

The familiar limit of the calculus, $\lim_{x \to a} f(x) = \ell$, is another instance of a net limit. Here D consists of the points x near a and $x \succ y$ means x is closer to a than y: $0 < |x - a| \leq |y - a|$.

Problem 1. Describe D and $\prec$ so these limits are limits of nets:

(i) $\lim_{x\to\infty} f(x) = \ell$,

(ii) $\lim_{x\to a+} f(x) = \ell$,

(iii) $\sum_{i=1}^{\infty} a_i = \ell$.

Proposition 1. *Let* $\{x_\alpha\}$ *and* $\{y_\alpha\}$ *be real valued nets on the same directed set D, with* $\lim_\alpha x_\alpha = \ell$, $\lim_\alpha y_\alpha = m$. *Then*

(i) $\lim_\alpha (x_\alpha + y_\alpha) = \ell + m$;

(i) $\lim_\alpha (x_\alpha - y_\alpha) = \ell - m$;

(iii) $\lim_\alpha (x_\alpha y_\alpha) = \ell m$;

(iv) $\lim_\alpha (x_\alpha / y_\alpha) = \ell / m$ *if* $m \neq 0, y_\alpha \neq 0$ *for all* α.

Proof. The proofs of these statements, which are basically just familiar properties of addition and multiplication, are virtually the same as the corresponding statements for sequences or functions. We prove (iii) by way of illustration.

Assume $x_\alpha \longrightarrow \ell$ and $y_\alpha \longrightarrow m$, or equivalently, $x_\alpha - \ell \longrightarrow 0$, $y_\alpha - m \longrightarrow 0$. Let $r_\alpha = x_\alpha - \ell$ and $s_\alpha = y_\alpha - m$ for all α. Then $r_\alpha \longrightarrow 0$ and $s_\alpha \longrightarrow 0$ and

$$\begin{aligned} x_\alpha y_\alpha &= (\ell + r_\alpha)(m + s_\alpha) \\ &= \ell m + r_\alpha m + s_\alpha \ell + r_\alpha s_\alpha . \end{aligned}$$

Fix $\varepsilon > 0$ and pick α_1, beyond which $|r_\alpha| < 1$ and in addition $|r_\alpha|$ is so small that $|r_\alpha m| < \varepsilon$. We similarly pick α_2 so that beyond α_2, $|s_\alpha| < \varepsilon$ and $|s_\alpha \ell| < \varepsilon$. There is α_0 so that $\alpha_0 \succ \alpha_1$ and $\alpha_0 \succ \alpha_2$. Hence if $\alpha \succ \alpha_0$, we have

$$|r_\alpha m| < \varepsilon, \quad |r_\alpha| < 1, \quad |s_\alpha \ell| < \varepsilon, \quad |s_\alpha| < \varepsilon ,$$

so

$$\begin{aligned} |x_\alpha y_\alpha - \ell m| &= |r_\alpha m + s_\alpha \ell + r_\alpha s_\alpha| \\ &\le |r_\alpha m| + |s_\alpha \ell| + |r_\alpha s_\alpha| < 3\varepsilon . \end{aligned}$$

Problem 2. Prove parts (i) and (iv) of Proposition 1.

Problem 3. (i) If $x_\alpha \ge 0$ for all α in a directed set D and $x_\alpha \longrightarrow \ell$, then $\ell \ge 0$.

(ii) If $x_\alpha \le y_\alpha \le z_\alpha$ for all $\alpha \in D$, and $x_\alpha \longrightarrow \ell$, $z_\alpha \longrightarrow \ell$, then $y_\alpha \longrightarrow \ell$.

Problem 4. Let D be the set of all pairs (m,n) of positive integers. Partially order D as follows: $(m,n) \succ (m',n')$ if and only if $m+n \ge m'+n'$.

(i) Describe geometrically what $(m,n) \succ (m',n')$ means.

(ii) Show that if $\lim\limits_{(m,n)} x_{m,n} = \ell$, then $\lim\limits_{n\to\infty} x_{m,n} = \ell$ for all m, and $\lim\limits_{m\to\infty} x_{m,n} = \ell$ for all n.

(iii) Let $x_{m,n} = mn/(m^2+n^2)$. Show $\lim\limits_{m\to\infty} x_{m,n} = 0$ for all n and $\lim\limits_{n\to\infty} x_{m,n} = 0$ for all m but $\lim\limits_{(m,n)} x_{m,n}$ fails to exist.

Problem 5. Let D be the set of all pairs (m,n) of natural numbers, with the partial ordering $(m,n) \succ (m_0,n_0)$ iff $\max\{m,n\} \ge \max\{m_0,n_0\}$. (i) Describe geometrically the set of (m,n) such that $(m,n) \succ (m_0,n_0)$ for a fixed (m_0,n_0).

(ii) Does $\lim\limits_{(m,n)} x_{m,n} = \ell$ imply $\lim\limits_{n\to\infty} x_{m,n} = \ell$ for all m?

Problem 6. Let D be as above with the ordering $(m,n) \succ (m',n')$ iff $mn \ge m'n'$. Give examples of nets $\{x_{m,n}\}$ which converge and nets which diverge. What is the connection, if any, between convergence in the ordering of D and the limits $\lim\limits_{m\to\infty} x_{m,n}$, $\lim\limits_{n\to\infty} x_{m,n}$?

The nets we want to consider here, and later for the Lebesgue integral, are nets of Riemann sums. Again let f be any real function on $[a,b]$ and let $P = \{x_0, x_1, \ldots, x_n\}$ be any partition of $[a,b]$. A **choice function** c for the partition P is a finite sequence $c_1, c_2, \ldots, c_n$ with $c_i \in [x_{i-1}, x_i)$ for $i < n$ and $c_n \in [x_{n-1}, x_n]$. We remind the reader that this notational clumsiness is deliberate so that the subintervals will be disjoint. The **Riemann sum** for f, P, and c is

$$R(f,P,c) = \sum_{i=1}^{n} f(c_i)(x_i - x_{i-1}).$$

The Riemann sum $R(f,P,c)$, for fixed f, is a real valued function which depends on the partition P and the choice function c. Thus

$\{R(f,P,c)\}$ becomes a net when we put an appropriate partial ordering $\prec$ on the pairs (P,c). We do this as follows:

$$(P_1, c_1) \succ (P_2, c_2) \text{ iff } P_1 \succ P_2 .$$

Thus the pairs are ordered by the partitions themselves in the sense that (P_1, c_1) is farther out than (P_2, c_2) if P_1 is a refinement of P_2.

We say that a net $\{x_\alpha\}$ is **increasing** provided $x_\beta \geq x_\alpha$ whenever $\beta \succ \alpha$. A **decreasing** net is defined similarly. The net $\{x_\alpha\}$ is **bounded** provided there are numbers b and B so that $b \leq x_\alpha \leq B$ for all α.

Problem 7. Show that an increasing bounded net $\{x_\alpha\}$ converges to $\ell = \sup\{x_\alpha : \alpha \in D\}$.

The lower sums $L(f,P)$ and the upper sums $U(f,P)$ for a bounded function are nets, with the partitions ordered by refinement. If f is the function in question and $m \leq f(x) \leq M$, then the lower sums and upper sums are bounded. The lower sums form an increasing bounded net, and hence converge, and similarly for the upper sums. Clearly f is Riemann integrable if and only if

$$\lim_P L(f,P) = \lim_P U(f,P). \tag{2}$$

If (P,c) is a partition of $[a,b]$ with a choice function c, then

$$m_i \leq f(c_i) \leq M_i$$

for each i, so

$$L(f,P) \leq R(f,P,c) \leq U(f,P). \tag{3}$$

It follows immediately from (2), (3) that if f is integrable, then

$$\lim_P R(f,P,c) = \int_a^b f. \tag{4}$$

We want to show that the existence of the limit $\lim_P R(f,P,c)$ provides an alternative characterization of integrability. Notice however, that to talk of lower and upper sums we need to assume that f is bounded. The Riemann sums, on the other hand, are defined even if f is not bounded, as long as f is defined on all of $[a,b]$. Conceivably the limit in (4) could exist for an unbounded function f, and we would have two distinct definitions of the integral. The next proposition resolves this question.

Proposition 2. *If f is defined on $[a,b]$ and $\lim_P R(f,P,c)$ exists, then f is bounded.*

Proof. Let $\lim_P R(f,P,c) = I$ and assume that f is not bounded above. Let P be a partition of $[a,b]$ such that for all choices c,

$$|\, R(f,P,c) - I \,| < 1\,.$$

In particular $|\, R(f,P,c) - R(f,P,c') \,| < 2$ for any choices c and c' for P. Since f is unbounded, f is unbounded on some subinterval, which we will assume is $[x_0, x_1)$. Fix any choice c for P. Let $c_i' = c_i$ for $i = 2, \ldots, n$, and choose c_1' so $f(c_1') > N$. Then

$$\begin{aligned} R\,(f,P,c') - R(f,P,c) &= (f\,(c_1') - f(c_1))\,\Delta x_1 \\ &> (N - f(c_1))\Delta x_1\,. \end{aligned}$$

We can choose N so large that the two Riemann sums differ by more than 2, which is a contradiction. ▮

Proposition 3. *A function f is integrable on $[a,b]$ if and only if $\lim_P R(f,P,c)$ exists, and of course the limit is the integral in this case.*

Proof. We have only to show that the existence of the limit implies that the function is integrable. To do this, fix $\varepsilon > 0$ and choose a partition P so that $|\; R(f,P,c) - I \;| < \varepsilon$ for all choices c, which implies that for any two choices c, c' for this P, $|\, R(f,P,c) - R(f,P,c') \,| < 2\varepsilon$. We will choose c and c' so that

$$L(f,P) \le R(f,P,c) < L(f,P) + \varepsilon\,, \tag{5}$$

$$U(f,P) - \varepsilon < R(f,P,c') \le U(f,P)\,. \tag{6}$$

This implies that $U(f,P) - L(f,P) < 4\,\varepsilon$. To see how c and c' can be chosen to satisfy (5) and (6), let

$$P = \{x_0, x_1, \ldots, x_n\}.$$

For each i, choose c_i so that

$$m_i \le f(c_i) < m_i + \varepsilon/(b-a).$$

Then

$$\begin{aligned}
L(f,P) &= \sum m_i \Delta x_i \\
&\le \sum f(c_i)\Delta x_i = R(f,P,c) \\
&< \sum \left(m_i + \frac{\varepsilon}{(b-a)} \right) \Delta x_i \\
&= \sum m_i \Delta x_i + \frac{\varepsilon}{b-a} \sum \Delta x_i \\
&= L(f,P) + \varepsilon \,.
\end{aligned}$$

We similarly choose c' so that for each i,

$$M_i \ge f\,(c_i') \ge M_i - \varepsilon/(b-a)\,,$$

and we have (6) by the same argument as above. ▮

Proposition 4. *If f and g are integrable on $[a,b]$ and k is a constant, then (i) $\int_a^b (f+g) = \int_a^b f + \int_a^b g$;*
(ii) $\int_a^b kf = k \int_a^b f$;
(iii) $\int_a^b f \ge 0$ if $f \ge 0$.

Proof. For every partition–choice pair (P,c),

$$\begin{aligned}
R(f+g,P,c) &= R(f,P,c) + R(g,P,c), \\
R(kf,P,c) &= kR(f,P,c).
\end{aligned}$$

Hence (i) and (ii) follow from the general results for nets in Proposition 1. Similarly, $R(f,P,c) \ge 0$ for all (P,c) if $f \ge 0$, so $\lim_P R(f,P,c) \ge 0$. ▮

Problem 8. If F is a continuous function on $\mathbb{R}$ and $\{x_\alpha\}$ is a net such that $x_\alpha \longrightarrow x$, then $\{F(x_\alpha)\}$ is a convergent net and $F(x_\alpha) \longrightarrow F(x)$. As a special case, if $x_\alpha \longrightarrow x$ then $|x_\alpha| \longrightarrow |x|$. Apply this to show that if f and hence $|f|$ are integrable, so that $\int_a^b f = \lim_P R(f,P,c)$ and $\int_a^b |f| = \lim_P R(|f|,P,c)$, then $|\int_a^b f| \le \int_a^b |f|$. ◀

Proposition 5. *If f is integrable on $[a,b]$ and F is a continuous function on $[a,b]$, and differentiable on (a,b) with $F'(x) = f(x)$ on (a,b), then*

$$\int_a^b f = F(b) - F(a).$$

Proof. Let $\varepsilon > 0$ and let P_0 be a partition such that $|R(f,P,c) - \int_a^b f| < \varepsilon$ whenever $P \succ P_0$ and c is any choice function for P. That is,

$$\left| \sum f(c_i)(x_i - x_{i-1}) - \int_a^b f \right| < \varepsilon,$$

$$\left| \sum F'(c_i)(x_i - x_{i-1}) - \int_a^b f \right| < \varepsilon, \tag{7}$$

whenever $P \succ P_0$ and c is any choice for P. The hypotheses for the ordinary Mean Value Theorem for F hold on each interval $[x_{i-1}, x_i]$. Therefore there is $c_i \in (x_{i-1}, x_i)$ for each i so that

$$\begin{aligned} F(x_i) - F(x_{i-1}) &= F'(c_i)(x_i - x_{i-1}) \\ &= f(c_i)(x_i - x_{i-1}). \end{aligned} \tag{8}$$

Let c be a choice function for P such that (8) holds. Then from (7) and (8) we get

$$\left| \sum_{i=1}^{n} (F(x_i) - F(x_{i-1})) - \int_a^b f \right| < \varepsilon.$$

Since $\sum (F(x_i) - F(x_{i-1})) = F(b) - F(a)$,

$$\left| F(b) - F(a) - \int_a^b f \right| < \varepsilon.$$

Since ε is arbitrary, equality holds. ▮

Consider the problem of summing an arbitrary collection of numbers. Say a little boy hands you a basket of numbers and asks you to add them - what do you do? You empty the numbers out on the floor, kick them into a row, and start adding from left to right. If there is a finite number of numbers, then there is no problem. If there is an infinite number of numbers in the basket, then you keep adding from left to right until you determine a limit, and that is the sum. The difficulty with this process is that if you sweep up all the numbers, put them back in the basket, and repeat the process, you will likely get a different answer. Indeed, unless all but a finite number of the numbers have the same sign you will almost surely get different answers on your second and subsequent trials. The point is that a conditionally convergent series is a very artificial thing, unless you have some real reason to want the numbers to appear in a given order. The unordered sum defined next gives a more convincing generalization of finite-sum addition.

Let A be any "index set" of elements α, and let x_α be a real number for each $\alpha \in A$. For any finite subset $F \subset A$ define S_F to be the finite sum $\sum_{\alpha \in F} x_\alpha$. Partially order the finite subsets F of A by inclusion: $F_1 \succ F_2$ if $F_1 \supset F_2$. Then $\{S_F\}$ is a net on this partially ordered set. If $\{S_F\}$ converges to L we write $\sum_{\alpha \in A} x_\alpha = L$ and say the x_α are **summable**.

Problem 9. (i) Show that if $\lim_F S_F = L$ exists, then at most countably many x_α are non-zero; *i.e.*, there is a countable subset $C \subset A$ so that $x_\alpha = 0$ if $\alpha \notin C$. (Hint: Suppose first that all $x_\alpha \geq 0$. If uncountably many $x_\alpha > 0$ then there is n such that $x_\alpha \geq \frac{1}{n}$ for uncountably many α.) Observe that this shows that **countable** additivity is the most one can ever ask for. There is no such thing as (non-trivial) uncountable addition.

(ii) Show that if $\sum_{\alpha \in A} x_\alpha = L$, then the set of positive x_α is summable and the set of negative x_α is summable and

$$\sum_{\alpha \in A^+} x_\alpha + \sum_{\alpha \in A^-} x_\alpha = \sum_{\alpha \in A} x_\alpha$$

where $A^+ = \{\alpha : x_\alpha > 0\}$, $A^- = \{\alpha : x_\alpha < 0\}$.

(iii) If $\{x_\alpha : \alpha \in A\}$ is summable, then $\{|x_\alpha| : \alpha \in A\}$ is summable and

$$\sum_{\alpha\in A} |x_\alpha| = \sum_{\alpha\in A^+} x_\alpha - \sum_{\alpha\in A^-} x_\alpha .$$

(iv) If $\sum_{n=1}^{\infty} x_n$ is absolutely convergent, then $\{x_n : n \in \mathbb{N}\}$ is summable, and conversely. In either case, $\sum_{n=1}^{\infty} x_n = \sum_{n\in\mathbb{N}} x_n$. ◀

Let $\{x_\alpha\}$ be a net on the directed set D. We will say that $\{x_\alpha\}$ is a **Cauchy net** provided that for each $\varepsilon > 0$ there is $\alpha_0 \in D$ so that $|x_\beta - x_\gamma| < \varepsilon$ whenever $\beta \succ \alpha_0$ and $\gamma \succ \alpha_0$.

Proposition 6. *If $\{x_\alpha\}$ is a Cauchy net then $\{x_\alpha\}$ converges.*

Proof. For each n pick α_n so that $|x_\beta - x_\gamma| < \frac{1}{n}$ when $\beta, \gamma \succ \alpha_n$. We can assume that $\alpha_1 \prec \alpha_2 \prec \alpha_3 \prec \cdots$ by replacing α_2 if necessary by α_2' with $\alpha_2' \succ \alpha_2$, $\alpha_2' \succ \alpha_1$, and α_3 by α_3' with $\alpha_3' \succ \alpha_3$ and $\alpha_3' \succ \alpha_2'$, etc. Then $\{x_{\alpha_n}\}$ is a Cauchy sequence, so there is a limit ℓ, and given $\varepsilon > 0$ there is N with $| x_{\alpha_N} - \ell |< \varepsilon$. We can assume $\frac{1}{N} < \varepsilon$. Then if $\beta \succ \alpha_N$, $|x_\beta - x_{\alpha_N}| < \varepsilon$ and $| x_{\alpha_N} - \ell |< \varepsilon$ so $| x_\beta - \ell |< 2\varepsilon$ if $\beta \succ \alpha_N$. ∎

Problem 10. Every convergent net is a Cauchy net. ◀

Problem 11. If A is an index set and $\{x_\alpha : \alpha \in A\}$ is summable, and $A = \bigcup A_n$ where the A_n are disjoint subsets of A, then $\{x_\alpha : \alpha \in A_n\}$ is summable for each n and $\sum_{\alpha\in A} x_\alpha = \sum_{n\in\mathbb{N}} \sum_{\alpha\in A_n} x_\alpha$. ◀

Problem 12. (i) If $\{x_{mn}\}$ is summable over $\mathbb{N} \times \mathbb{N}$, then $\sum_{(m,n)} x_{mn} = \sum_{m=1}^{\infty} \sum_{n=1}^{\infty} x_{mn}$.

(ii) Suppose the iterated (ordered) sums both exist. Does it follow that $\{x_{mn}\}$ is summable?

(iii) Suppose the iterated sums both exist and $x_{mn} \geq 0$ for all m, n. Does it follow that $\{x_{mn}\}$ is summable? ◀

Problem 13. If $F(m,n) = \frac{1}{2}(m+n-2)(m+n-1)+n$, then F is a one-to-one function on $\mathbb{N}\times\mathbb{N}$ into $\mathbb{N}$. Hence the set of all pairs (m,n) is countable, and any countable union of countable sets is countable. Hint: If $f(x) = \frac{1}{2}(x-2)(x-1)$, then $F(m,n) = f(m+n)+n$. Show that $f(x+1) - f(x) = x - 1$, and conclude that if $m + n = i + j + 1$, $F(m,n) > F(i,j)$. Does F map $\mathbb{N} \times \mathbb{N}$ onto $\mathbb{N}$?

Problem 14. Let $\prec$ and $\olessthan$ be two partial orderings, both of which make D a directed set. Suppose $\alpha \prec \beta$ implies $\alpha \olessthan \beta$ for all $\alpha, \beta \in D$. Let $\{x_\alpha\}$ be a net on D, and let $\lim\limits_{\prec} x_\alpha$ and $\lim\limits_{\olessthan} x_\alpha$ denote the limits with respect to the two orderings. Show that if $\lim\limits_{\olessthan} x_\alpha = \ell$, then $\lim\limits_{\prec} x_\alpha = \ell$.

The next problem shows that the Riemann integral can be characterized as a limit of Riemann sums, where the partitions are not directed by refinement but by insisting that the length of the maximum subinterval tends to zero. We will use this result later to characterize the Riemann integrable functions as those which are continuous except on a set of measure zero.

Problem 15. If $P = \{x_0, \ldots, x_n\}$ is a partition of $[a,b]$, define the **norm** of P, denoted $\|P\|$, by $\|P\| = \max(x_i - x_{i-1})$. Let $P \olessthan Q$ mean that $\|Q\| \le \|P\|$. Show that $\olessthan$ makes the partitions P and the pairs (P,c) into a directed set. Let $\lim\limits_{\|P\| \to 0}$ stand for the limit with the direction $\olessthan$. Use Problem 14 to show that if $\lim\limits_{\|P\| \to 0} R(f,P,c) = I$, then f is Riemann integrable with integral I.

Conversely, $\lim\limits_{P} R(f,P,c) = I$ implies $\lim\limits_{\|P\| \to 0} R(f,P,c) = I$. Prove this. Hint: Let $P_0 = \{x_0, \ldots, x_n\}$ be a partition of $[a,b]$ such that $U(f,P_0) - L(f,P_0) < \varepsilon$, and so $|R(f,P_0,c_0) - I| < \varepsilon$. Let $Q = \{y_0, y_1, \ldots, y_k\}$ be another partition with $\|Q\| = \delta < \min(x_i - x_{i-1})$. Let J be the set of all indices j for Q such that $[y_{j-1}, y_j)$ is contained wholly within some $[x_{i-1}, x_i)$ of P. If M_j and m_j are the sups and infs of f for Q, then

$$\sum_{j \in J} (M_j - m_j) \triangle y_j \le U(f,P_0) - L(f,P_0).$$

If K denotes the Q indices not in J (so all j such that $[y_{j-1}, y_j)$ contains some x_i), then

$$\sum_{j \in K} (M_j - m_j) \triangle y_j < n(M - m)\delta$$

where M and m are bounds for f on $[a, b]$. Consequently, if $\|Q\| = \delta \leq \varepsilon/n(M - m)$, then

$$U(f, Q) - L(f, Q) \leq U(f, P_0) - L(f, P_0) + \varepsilon < 2\varepsilon,$$

and so $|R(f, Q, c) - I| < 2\varepsilon$. ◀

3 LEBESGUE MEASURE ON $(0, 1)$

Let f be the characteristic function of the rational numbers in $(0, 1)$; *i.e.*, the function which is one on the rationals and zero elsewhere. Then for any partition $P = \{x_0, x_1, \ldots, x_n\}$ of $[0, 1]$, $m_i = 0$ and $M_i = 1$ for all i. Here, as usual, m_i and M_i are the inf and sup of the function values on $[x_{i-1}, x_i)$. Since $L(f, P) = 0$ and $U(f, P) = 1$ for all partitions P, f is clearly not Riemann integrable.

Now recall the geometric interpretation of the integral as the area under the graph of f. If f is the characteristic function of the rationals in $[0, 1]$, then the region under the graph of f is a very simple one; it is just the "rectangle" $\mathbb{Q} \times I$ where $\mathbb{Q}$ is the set of rationals in $[0, 1]$ and $I = [0, 1]$. The area of this rectangle obviously ought to be the length of $\mathbb{Q}$ times the length of I. Once we have a sensible definition for the length of $\mathbb{Q}$ we will have a reasonable value for the integral of f.

In this chapter we extend the idea of length from intervals to all subsets of $\mathbb{R}$. This generalized length, called the (**Lebesgue outer**) **measure** of a set, will assign measure zero to $\mathbb{Q}$ (and all other countable sets) so that we will have no difficulty agreeing that $\int f = 0$ when f is the characteristic function of the rationals. The difference between the definitions of the Riemann and Lebesgue integrals consists in just this fact: for the Lebesgue integral we allow partitions into sets more general than intervals, and this requires that we can assign a length to these partitioning sets. With this one variation, the definition of the Lebesgue integral will be the same as the definition of the Riemann integral.

We now proceed with the definition of the measure function m. We restrict our attention initially to subsets of the open unit interval $U = (0, 1)$. This will ultimately give us also the measure of subsets of any interval $(n, n+1)$, since we want measure to be invariant under translation. Countable sets will turn out to have measure zero, so we will finally define the measure of any set E to be the sum of the measures of the sets $E \cap (n, n+1)$, $n = 0, \pm 1, \pm 2, \ldots$.

For any interval I we let $\ell(I)$ denote the length of I. An interval can be open, closed, or half-open, but not just a single point. Hence $\ell(I) > 0$ for every interval I. Roughly speaking, we define the measure of a set E to be the minimum of the sums of the lengths of families of intervals which cover E. To make this precise, we say a finite or countable family $\{I_j\}$ of intervals is a **covering of** E if $E \subset \cup I_j$. The family is an **open covering** if all the I_j are open intervals, and a **closed covering** if all the I_j are closed intervals. The **total length** of the family $\{I_j\}$ is $\sum \ell(I_j)$. Finally, we define the measure of E to be the infimum of the total lengths of all coverings of E:

$$m(E) = \inf\left\{\sum \ell(I_j) : E \subset \cup I_j\right\}.$$

Let us check that it does not matter in this definition whether we use open intervals or closed intervals or a mixture. For each open covering $\{I_j\}$ of E there is a closed covering $\left\{\bar{I}_j\right\}$ of the same total length, so using closed coverings might conceivably give a smaller value for $m(E)$. However, for each closed covering $\{I_k\}$ of E there is an open covering $\{J_k\}$ whose total length is less than $\sum \ell(I_k) + \varepsilon$. (Let J_k be an open interval containing I_k with $\ell(J_k) < \ell(I_k) + \varepsilon/2^k$.) Hence to each closed covering $\{I_j\}$ of E there are open coverings with total lengths arbitrarily close to the total length of $\{I_j\}$, so the infima of total lengths of coverings are the same.

If E is a compact set in $(0, 1)$ – *i.e.*, a closed bounded set – then every open covering of E has a finite subset which covers; this is the Heine–Borel Theorem. It follows that if E is compact, $m(E)$ is the inf of the total lengths of finite open coverings. By the discussion above, we could also use finite closed coverings to find $m(E)$ if E is compact. These remarks are perhaps unnecessarily complicated for defining measure on the line, since we could simply stick with coverings by open intervals. However, when we consider plane measure in Chapter 9 it will be convenient to know that the covering sets can be open or closed (rectangles), or anything in between.

The number $m(E)$ is the **Lebesgue outer measure** of E, and we will refer to $m(E)$ simply as the **measure** of E. Here are some immediate properties of m.

Proposition 1. *(i)* $0 \leq m(E) \leq 1$ *for all* $E \subset (0,1)$*;*
(ii) $m(E) \leq m(F)$ *if* $E \subset F$*; (*m *is monotone)*
(iii) $m(\emptyset) = 0$*;*
(iv) $m(\{x\}) = 0$ *for all* x*;*

Problem 1. Write out the very short proofs of parts (i), (ii), (iii), (iv) of Proposition 1. Note that subsets of sets of measure zero have measure zero. ◀

Since $m(E)$ is to be a generalization of length, we need to know that $m(I) = \ell(I)$ for every interval $I \subset (0,1)$. That is the content of the next proposition and problem.

Proposition 2. *If* $J = [a,b] \subset (0,1)$*, or if* $J = (a,b) \subset (0,1)$*, then* $m(J) = \ell(J)$*.*

Proof. First assume J is a closed interval $[a,b]$. Clearly $m(J) \leq \ell(J)$ since $\{J\}$ is a one-interval covering of J of total length $\ell(J)$. We show by induction that if $I_1, \ldots, I_n$ is any finite covering of J by intervals, then $\ell(J) \leq \sum_{k=1}^{n} \ell(I_k)$. If J is covered by one interval I_1, then clearly $\ell(J) \leq \ell(I_1)$. Suppose as our inductive hypothesis that whenever J is a closed interval covered by n or fewer open intervals $I_1, \ldots, I_m$ $(m \leq n)$ that $\ell(J) \leq \sum_{k=1}^{m} \ell(I_k)$. Let $J = [a,b] \subset I_1 \cup \cdots \cup I_{n+1}$, and assume no n of these intervals cover J. If any I_k is disjoint from J we are done. Let us assume, to be definite, that $I_{n+1} = (c,d)$ with $a < c < d < b$. Let $J_1 = [a,c]$ and $J_2 = [d,b]$ be the two subintervals of J not covered by I_{n+1}. No interval $I_k, k = 1, \ldots, n$, can intersect both J_1 and J_2 for such an interval would cover I_{n+1}, and n of the I_k would cover J. Therefore some of the intervals $I_1, \ldots, I_n$ cover J_1 and the rest cover J_2. By the inductive assumption applied separately to J_1 and J_2 we see that

$$\ell(J_1) + \ell(J_2) \le \sum_{k=1}^{n} \ell(I_k).$$

Since

$$\ell(J) = \ell(J_1) + \ell(I_{n+1}) + \ell(J_2),$$

we then have

$$\ell(J) \le \sum_{k=1}^{n+1} \ell(I_k).$$

This shows that $m(J) = \ell(J)$ for every closed interval. The cases where $I_{n+1} = (c, d)$ with $c \le a$ or $d \ge b$ are treated similarly. If $J = (a, b)$, then we pick a closed interval $[c, d] \subset (a, b)$ with $d - c > b - a - \varepsilon$. Hence, by monotonicity,

$$m(a, b) \ge m[c, d] = d - c > b - a - \varepsilon.$$

We already know that $m(I) \le \ell(I)$ for any I so $m(a, b) = \ell(a, b)$. The remaining cases are left as an exercise. ▮

Problem 2. Show that $m(0, 1) = 1$ and that $m(a, b] = m[a, b) = b - a$ for all half-open intervals in $(0, 1)$. ◂

Proposition 3. *m is countably subadditive; i.e., for any finite or countable family $\{E_i\}$ of subsets of $(0, 1)$,*

$$m(\cup E_i) \le \sum m(E_i).$$

Proof. Let $\varepsilon > 0$ and let $\{I_{ij}\}$ be a covering of E_i by open intervals so that

$$m(E_i) + \varepsilon/2^i > \sum_j \ell(I_{ij}).$$

Then $\cup E_i \subset \cup I_{ij}$ and hence

$$\begin{aligned} m(\cup E_i) &\le \sum_{ij} \ell(I_{ij}) \\ &= \sum_i \sum_j \ell(I_{ij}) \\ &\le \sum_i \left(m(E_i) + \frac{\varepsilon}{2^i} \right) \\ &= \sum_i m(E_i) + \varepsilon. \end{aligned}$$

Since this holds for all $\varepsilon > 0$,

$$m(\cup E_i) \le \sum m(E_i).$$

Problem 3. Show that countable sets have measure zero.

Problem 4. If $\mathbb{Q}$ is the set of rationals in $(0,1)$ then we know from Problem 3 that $m(\mathbb{Q}) = 0$, and hence for any $\varepsilon > 0$ there are open intervals $\{I_j\}$ so that $\mathbb{Q} \subset \cup I_j$ and $\sum \ell(I_j) < \varepsilon$. Show that if $\mathbb{Q} \subset I_1 \cup \cdots \cup I_n$ with $I_1, \ldots, I_n$ open intervals in $(0,1)$, then $\ell(I_1) + \cdots + \ell(I_n) \ge 1$. Hence, although finite coverings suffice to approximate $m(E)$ for compact sets E, arbitrary sets require countable coverings.

Problem 5. Let $E = E_1 \cup E_2$ with $m(E_2) = 0$. Show that $m(E) = m(E_1)$. Make precise and prove the assertion that if two sets differ by a set of measure zero, then the two sets have the same measure.

Problem 6. Let $E_1 \subset I_1$ and $E_2 \subset I_2$ where I_1 and I_2 are disjoint intervals in $(0,1)$. Show that $m(E_1 \cup E_2) = m(E_1) + m(E_2)$. Generalize to a finite number of sets $E_1, \ldots, E_n$.

Problem 7. (i) Let $E \subset (0,1)$, and let $E_r = \{x + r : x \in E\}$. If $E_r \subset (0,1)$, then $m(E) = m(E_r)$.

(ii) Let $r \in (0,1)$. For $x \in (0,1)$, define

$$x \oplus r = \begin{cases} x + r & \text{if} \quad x + r \in (0,1) \\ x + r - 1 & \text{if} \quad x + r > 1. \end{cases}$$

We need not define $x \oplus r$ if $x + r = 1$, since we want $x \oplus r$ to lie always in $(0,1)$. Let $E_r = \{x \oplus r : x \in E\}$, so that E_r is now the r-translate of E with the points that fall outside $(0,1)$ put back at the left end of the interval. Show that $m(E) = m(E_r)$.

Note. We will show in the next chapter that the measure function m is countably additive on a usefully large family of sets (the measurable sets) but not on all sets. To construct a non-measurable set we will need the kind of translation invariance in part (ii) above.

4 MEASURABLE SETS — THE CARATHÉODORY CHARACTERIZATION

The critical property of the measure function m is that it be additive. Ideally we should have an identity like

$$m\left(\bigcup E_i\right) = \sum m(E_i) \tag{1}$$

for all finite or countable disjoint families $\{E_i\}$. Unfortunately, m is **not** countably additive over all sets, and we must sort out the so-called measurable sets on which (1) does hold.

If E is any set in $(0,1)$ and E' is its complement in $(0,1)$, then a minimal requirement for additivity is certainly

$$m(E) + m(E') = m(0,1) = 1. \tag{2}$$

It turns out that this condition is sufficient to distinguish the sets E on which m is countably additive so that (1) holds. Hence we make the following definition:

A set $E \subset (0,1)$ is **measurable** if and only if

$$m(E) + m(E') = 1, \tag{3}$$

where $E' = (0,1) - E$.

It is immediate from the definition that E is measurable if and only if E' is measurable. Moreover, since m is subadditive, we automatically have

$$m(E) + m(E') \geq m(0,1) = 1.$$

Hence E is measurable if and only if

$$m(E) + m(E') \leq 1.$$

The measurable sets include the intervals, and are closed under countable unions and intersections. The measure m is additive on any finite or countable family of disjoint measurable sets. The verification of these facts is the program for this chapter.

A cautionary word about notation and nomenclature: most texts use m^* for our function m and refer to it as Lebesgue outer measure. The unadorned letter m is used by these authors for the restriction of m^* to the measurable sets, and this restricted function is called Lebesgue measure. We will stick to m, defined on all subsets of $(0,1)$, and call $m(E)$ the measure of E whether E is measurable or not. In practice (*i.e.*, following this chapter) we only consider $m(E)$ for measurable sets since measurability is essential for m to have the critical property of additivity.

Proposition 1. *(i) If $m(E) = 0$, E is measurable.*
(ii) Intervals are measurable.

Proof. (i) If $m(E) = 0$, then

$$m(E) + m(E') = m(E') \le 1\,,$$

and this inequality is equivalent to measurability.

(ii) Let $J = (a,b)$ be a proper subinterval of $(0,1)$ and let $J' = J_1 \cup J_2$ where J_1 and J_2 are the two complementary intervals to J. (One of J_1, J_2 will be empty if $J = (0,b)$ or $J = (a,1)$.) Since the measure of an interval is its length,

$$m(J_1) + m(J) + m(J_2) = 1.$$

Therefore,

$$\begin{aligned} m(J') &\le m(J_1) + m(J_2) = 1 - m(J), \\ m(J) + m(J') &\le 1. \end{aligned}$$

This argument works for closed and half-open intervals too. ▮

We saw in Problem 6 of the preceding chapter that if J_1 and J_2 are disjoint intervals in $(0,1)$, then for any set E,

$$m(E \cap (J_1 \cup J_2)) = m(E \cap J_1) + m(E \cap J_2).$$

We now extend this to finite or countable families $\{J_i\}$.

Proposition 2. *If $\{J_i\}$ is a finite or countable family of disjoint intervals in $(0,1)$, then for any set E,*

$$m\left(E \cap \bigcup J_i\right) = \sum m(E \cap J_i)\,.$$

Proof. First consider a finite family $J_1, \ldots, J_n$ of disjoint intervals, and assume without loss that $E \subset J_1 \cup \cdots \cup J_n$. Let $\{I_k\}$ be a covering of E such that

$$\sum \ell(I_k) < m(E) + \varepsilon\,.$$

The sets $I_k \cap J_\ell$, $k = 1, 2, \cdots$, form a covering of $E \cap J_\ell$. Moreover,

$$\ell(I_k \cap J_1) + \cdots + \ell(I_k \cap J_n) \leq \ell(I_k)$$

by Problem 6 of the preceding chapter. Hence

$$m(E \cap J_\ell) \leq \sum_k \ell(I_k \cap J_\ell)\,,$$

and

$$\begin{aligned} m(E) &\leq \sum_\ell m(E \cap J_\ell) \\ &\leq \sum_\ell \sum_k \ell(I_k \cap J_\ell) \\ &= \sum_k \sum_\ell \ell(I_k \cap J_\ell) \\ &\leq \sum_k \ell(I_k) \\ &< m(E) + \varepsilon\,. \end{aligned}$$

Therefore $m(E) = \sum m(E \cap J_\ell)$ if $E \subset J_1 \cup \cdots \cup J_n$. Now let $\{J_k\}$ be a countable disjoint covering of E by intervals. Then using subadditivity in the first inequality below and monotonicity in the last we get

$$
\begin{aligned}
m\left(E \cap \bigcup_{k=1}^{\infty} J_k\right) &\le \sum_{k=1}^{\infty} m(E \cap J_k) \\
&= \lim_{n\to\infty} \sum_{k=1}^{n} m(E \cap J_k) \\
&= \lim_{n\to\infty} m(E \cap (J_1 \cup \cdots \cup J_n)) \\
&\le m\left(E \cap \bigcup_{k=1}^{\infty} J_k\right).
\end{aligned}
$$

Hence the first inequality is an equality and we are done. ▮

The next result is a formalization of the statement that measurability is a local property.

Proposition 3. *E is measurable if and only if for every interval* $J \subset (0,1)$,

$$m(E \cap J) + m(E' \cap J) = m(J).$$

Proof. Suppose, for example, that $J = (a,b)$ with $0 < a < b < 1$. Let $J_1 = (0,a)$, $J_2 = J = (a,b)$, $J_3 = (b,1)$. Since the two element set $\{a,b\}$ has measure zero, $m(E \sim \{a,b\}) = m(E)$ and similarly for E', so

$$
\begin{aligned}
m(E) &= m(E \cap J_1) + m(E \cap J_2) + m(E \cap J_3), \\
m(E') &= m(E' \cap J_1) + m(E' \cap J_2) + m(E' \cap J_3).
\end{aligned}
$$

Adding columnwise we get

$$
\begin{aligned}
m(E) + m(E') &= \sum_{i=1}^{3} \Big[m(E \cap J_i) + m(E' \cap J_i) \Big] \\
&\ge \sum_{i=1}^{3} m(J_i) = 1.
\end{aligned}
$$

Hence $m(E) + m(E') = 1$ if and only if

$$m(E \cap J_i) + m(E' \cap J_i) = m(J_i)$$

for each i. If J is of the form $(0, b)$ or $(a, 1)$ the same argument works by considering the single complementary interval. ▮

Problem 1. Carry out the proof of Proposition 3 in the case $J = (0, b)$. ◀

Our definition says E is measurable if E splits $(0, 1)$ additively. Proposition 3 shows that E is measurable only if E splits every subinterval of $(0, 1)$ additively. We next show that E is measurable if and only if E splits **every** subset T additively. This result is due to Carathéodory and has become the modern **definition** of measurability in all general settings. That is, given any countably subadditive non–negative monotone function m defined on all subsets of **any** set, the function m will be countably additive when restricted to the sets E which satisfy

$$m(E \cap T) + m(E' \cap T) = m(T)$$

for all subsets T.

Proposition 4. *E is measurable if and only if for every "test set" $T \subset (0, 1)$,*

$$m(E \cap T) + m(E' \cap T) = m(T)\,.$$

Since m is subadditive, E is measurable if and only if for every set T

$$m(E \cap T) + m(E' \cap T) \le m(T)\,. \tag{4}$$

Proof. The condition (4) is clearly sufficient since we can take $T = (0, 1)$. To show that (4) holds for every measurable set E, let T be any set in $(0, 1)$, $\varepsilon > 0$, and $\{I_j\}$ a covering of T by open intervals such that

$$\sum m(I_j) < m(T) + \varepsilon\,.$$

Then

$$E \cap T \subset \bigcup (E \cap I_j), \quad E' \cap T \subset \bigcup (E' \cap I_j)\,.$$

If E is measurable, then

$$m(E \cap I_j) + m(E' \cap I_j) = m(I_j) \tag{5}$$

for each j. Hence, using monotonicity, subadditivity, and (5),

$$m(E \cap T) + m(E' \cap T) \leq \sum m(E \cap I_j) + \sum m(E' \cap I_j)$$
$$= \sum m(I_j) < m(T) + \varepsilon\,.$$

Since ε is arbitrary we have the desired inequality (4). ▮

Problem 2. If E_1, E_2 are measurable sets then $m(E_1 - E_2) = m(E_1) - m(E_1 \cap E_2)$ and $m(E_1 \cup E_2) = m(E_1) + m(E_2) - m(E_1 \cap E_2)$. Do you need all the hypotheses? ◀

We show next that m is countably additive on measurable sets, and that the measurable sets are closed under countable unions and intersections. The discerning reader will notice, with Carathéodory, that the next three propositions make no use of the fact that the sets are subsets of $(0, 1)$, or of how the function m is defined. We use just these facts:

$$m(\varnothing) = 0\,,$$
$$0 \leq m(E) \leq \infty\,,$$
$$m(E) \geq m(F) \text{ if } E \supset F\,,$$
$$m\left(\bigcup E_i\right) \leq \sum m(E_i)\,,$$

and for measurable sets E, and all T,

$$m(E \cap T) + m(E' \cap T) = m(T)\,. \tag{6}$$

Proposition 5. *If $\{E_i\}$ is a finite or countable family of disjoint measurable sets, then*

$$m\left(\bigcup E_i\right) = \sum m(E_i)\,.$$

Proof. Let $E_1, \ldots, E_n$ be disjoint measurable sets. The set E_1 cuts the test set $T = E_1 \cup \cdots \cup E_n$ additively, so

$$m(E_1) + m(E_2 \cup \cdots \cup E_n) = m(E_1 \cup \cdots \cup E_n)\,.$$

The measurable set E_2 cuts $E_2 \cup \cdots \cup E_n$ additively, so

$$m(E_2) + m(E_3 \cup \cdots \cup E_n) = m(E_2 \cup \cdots \cup E_n)\,,$$

and hence

$$m(E_1) + m(E_2) + m(E_3 \cup \cdots \cup E_n) = m(E_1 \cup \cdots \cup E_n) .$$

In a finite number of steps we have

$$m\left(\bigcup_{i=1}^{n} E_i\right) = \sum_{i=1}^{n} m(E_i) .$$

Now let $\{E_i\}$ be a countable family of disjoint measurable sets. For each n,

$$m\left(\bigcup_{i=1}^{\infty} E_i\right) \geq m\left(\bigcup_{i=1}^{n} E_i\right) = \sum_{i=1}^{n} m(E_i) ,$$

and hence

$$m\left(\bigcup_{i=1}^{\infty} E_i\right) \geq \sum_{i=1}^{\infty} m(E_i) .$$

The opposite inequality is automatic by subadditivity, so equality holds. ▮

Problem 3. Show that if $\{E_i\}$ is a countable disjoint family of measurable sets and T is any set, then

$$m\left(T \cap \bigcup E_i\right) = \sum m(T \cap E_i) . \quad ◀$$

Proposition 5 tells us what $m\left(\bigcup E_i\right)$ is if the E_i are measurable (and pairwise disjoint), but does not say that $\bigcup E_i$ is itself measurable. This we show next, proving first that $E_1 \cup E_2$ is measurable.

Proposition 6. *If E_1 and E_2 are measurable, then $E_1 \cup E_2$ is measurable.*

Proof. Let E_1 and E_2 be measurable sets and let T be any test set. Let $T = T_1 \cup T_2 \cup T_3 \cup T_4$ as indicated in Fig. 1 (next page).

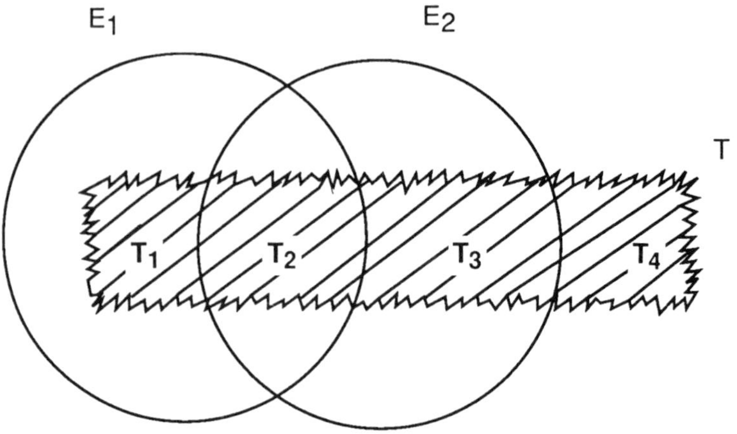

Fig. 1

What we must show is

$$m[(E_1 \cup E_2) \cap T] + m[(E_1 \cup E_2)' \cap T] = m(T);$$

or, in terms of Fig. 1,

$$m(T_1 \cup T_2 \cup T_3) + m(T_4) = m(T).$$

Cutting the test set $T_1 \cup T_2$ with the measurable set E_2 gives

$$m(T_1) + m(T_2) = m(T_1 \cup T_2). \tag{7}$$

Similarly, cutting $T_3 \cup T_4$ with E_2 gives

$$m(T_3) + m(T_4) = m(T_3 \cup T_4). \tag{8}$$

Cutting T with E_1 gives

$$m(T_1 \cup T_2) + m(T_3 \cup T_4) = m(T). \tag{9}$$

Combining (7), (8), (9) we can write

$$m(T_1) + m(T_2) + m(T_3) + m(T_4) = m(T). \tag{10}$$

Now cut $T_1 \cup T_2 \cup T_3$ with E_1 and then use (7):

$$m(T_1 \cup T_2) + m(T_3) = m(T_1 \cup T_2 \cup T_3),$$

$$m(T_1) + m(T_2) + m(T_3) = m(T_1 \cup T_2 \cup T_3). \tag{11}$$

From (11) and (10) we have the desired equality

$$m(T_1 \cup T_2 \cup T_3) + m(T_4) = m(T). \quad ||||$$

Corollary. *Finite unions and finite intersections of measurable sets are measurable.* $E_1 - E_2$ *is measurable if* E_1, E_2 *are.*

Proof. Notice that E satisfies the characterizing equation

$$m(E \cap T) + m(E' \cap T) = m(T)$$

for all T if and only if E' does. Therefore

$$E_1 \cap E_2 = (E_1' \cup E_2')'$$

is measurable whenever E_1, E_2 are. The inductive proof from two sets to a finite number is immediate. Since

$$E_1 - E_2 = E_1 \cap E_2' ,$$

differences are measurable. ▮

Proposition 7. *If* $\{E_i\}$ *is a countable family of measurable sets, then* $\cup E_i$ *is measurable and* $\cap E_i$ *is measurable. Open sets and closed sets are measurable.*

Proof. We can assume the E_i are disjoint by replacing E_2 by $E_2 - E_1$, E_3 by $E_3 - (E_1 \cup E_2)$, etc. Let $F_n = E_1 \cup \cdots \cup E_n$, so F_n is measurable and by Proposition 5

$$m(F_n) = \sum_{i=1}^{n} m(E_i) .$$

For any test set T, by Problem 3,

$$\begin{aligned} m(T) &= m(T \cap F_n) + m(T \cap F_n') \\ &= \sum_{i=1}^{n} m(T \cap E_i) + m(T \cap F_n') . \end{aligned}$$

If $E = \bigcup E_i$, then $F_n \subset E$ for all n so $F_n' \supset E'$, $m(T \cap F_n') \geq m(T \cap E')$ and

$$m(T) \geq \sum_{i=1}^{n} m(T \cap E_i) + m(T \cap E') .$$

This last inequality holds for all n, so, again by Problem 3,

$$\begin{aligned} m(T) &\geq \sum_{i=1}^{\infty} m(T \cap E_i) + m(T \cap E') \\ &= m(T \cap E) + m(T \cap E'), \end{aligned}$$

and $E = \bigcup_{i=1}^{\infty} E_i$ is measurable.

The remainder is left as a problem. ▮

A σ-**algebra** of subsets of any set X is a family of subsets which contains X and $\emptyset$, and is closed under finite or countable unions, finite or countable intersections, and complementation. Since $E_1 - E_2 = E_1 \cap E_2'$, σ-algebras are closed under differences.

Problem 4. Show that the measurable subsets of $(0, 1)$ form a σ-algebra, and that every open or closed set is measurable. ◀

Problem 5. Let E be a measurable subset of $(0, 1)$. Show that for each $\varepsilon > 0$ there is an open set U and a closed set F such that $F \subset E \subset U$ and

$$m(E) - \varepsilon \leq m(F) \leq m(U) < m(E) + \varepsilon.$$

Show that the existence of such F and U for every $\varepsilon > 0$ is also sufficient for E to be measurable. ◀

Now we extend the definition of Lebesgue measure to include arbitrary subsets of $\mathbb{R}$. We use μ for the extended measure function, so μ is defined on all subsets of $\mathbb{R}$. For a set $E \subset (n, n+1)$ define $\mu(E) = m(E - n)$, where $E - n = \{x - n : x \in E\}$. If $E \subset (0, 1)$, then of course $\mu(E) = m(E)$. For any set $E \subset \mathbb{R}$, define

$$\mu(E) = \sum_{n=-\infty}^{\infty} \mu(E \cap (n, n+1)).$$

If E contains integer points, that will not affect the value of $\mu(E)$ since we still want countable sets to have measure zero. Notice that $\mu(E) = \infty$ is now a possibility.

We will say that E is measurable if and only if $E \cap (n, n+1)$ is measurable for each n; *i.e.*, if

$$\mu(E \cap (n, n+1)) + \mu(E' \cap (n, n+1)) = 1$$

for each n. We could alternatively have used the Carathéodory criterion to define measurability, as the next proposition shows.

Proposition 8. *A subset $E \subset \mathbb{R}$ is measurable iff for every set $T \subset \mathbb{R}$,*

$$\mu(E \cap T) + \mu(E' \cap T) = \mu(T) . \tag{12}$$

Proof. By definition,

$$\begin{aligned}\mu(T) &= \sum_{n=-\infty}^{\infty} \mu(T \cap (n, n+1)) , \\ \mu(T \cap E) &= \sum_{n=-\infty}^{\infty} \mu(T \cap E \cap (n, n+1)) , \\ \mu(T \cap E') &= \sum_{n=-\infty}^{\infty} \mu(T \cap E' \cap (n, n+1)) .\end{aligned}$$

If E is measurable then

$$\mu(T \cap E \cap (n, n+1)) + \mu(T \cap E' \cap (n, n+1)) = \mu(T \cap (n, n+1))$$

for each n, and addition gives (12). On the other hand, if (12) holds for all T, then we see that each $E \cap (n, n+1)$ is measurable by letting $T = (n, n+1)$. ▮

Problem 6. Show that μ is translation invariant; *i.e.*, if $E + x = \{y + x : y \in E\}$, then $\mu(E) = \mu(E + x)$ for all E and x. ▮

We finish by constructing a non-measurable subset E of $(0, 1)$. To show that E is not measurable we show that $(0, 1)$ is the union of a countable number of disjoint translates (modulo 1) of E. If E were measurable and $m(E) = 0$, then we could conclude that $m(0, 1) = \sum_{n=1}^{\infty} m(E) = 0$. If $m(E) > 0$, then we could conclude that $m(0, 1) = \sum_{n=1}^{\infty} m(E) = \infty$.

The following "construction" of a non-measurable set E depends heavily on the Axiom of Choice, or some equivalent logical assumption such as Zorn's Lemma. Although some mathematicians pretend to be frightened by the Axiom of Choice, we will cheerfully accept it here as part of our common logic.

Pick any $x \in (0,1)$, and then any y so $y - x$ is not rational. Then pick z so $z - x$ and $z - y$ are not rational. Continue this process - uncountably many times - until there remains no number in $(0,1)$ which is not obtained by adding a rational to one of the already chosen numbers. The set E is thus a maximal subset of $(0,1)$ with the property that all differences $x - y$ for $x, y \in E$ are irrational. Hence for all $t \notin E$, $t = x + r$ for some $x \in E$ and some rational r. Let $r_1, r_2, \ldots$ be an enumeration of the rationals in $[0,1)$ and let E_n consist of all numbers $x + r_n$ (modulo 1) for $x \in E$. That is, if $x \in E$, $x + r_n \in E_n$ if $x + r_n < 1$ and $x + r_n - 1 \in E_n$ if $x + r_n > 1$. Since the sets E_n are essentially just translates of E all E_n have the same measure. (See Problem 7, Chapter 3.) The E_n are disjoint, for if $x, y \in E$ and

$$\left.\begin{matrix} x + r_n \\ \text{or} \\ x + r_n - 1 \end{matrix}\right\} = \left\{\begin{matrix} y + r_m \\ \text{or} \\ y + r_m - 1 \end{matrix}\right.$$

then $x - y$ is rational. Clearly $(0,1) = \bigcup E_n$ since every t not in E has the form $x + r_n$ (modulo 1) for some $x \in E$, some rational r_n. Thus $(0,1)$ is a countable union of disjoint sets with the same measure. If E is measurable then all E_n are measurable and $m(0,1) = \sum m(E_n)$, which is zero or infinity.

Problem 7. Use the Carathéodory characterization (6) to show that if for every $\varepsilon > 0$ there are measurable sets A and B such that $A \subset E \subset B$ and $\mu(B) - \mu(A) < \varepsilon$, then E is measurable.

Problem 8. (i) If $\{E_i\}$ is a sequence of measurable sets such that $\mu(E_1) < \infty$ and $E_1 \supset E_2 \supset E_3 \supset \cdots$, then $\mu\left(\bigcap E_i\right) = \lim_i \mu(E_i)$. Hint: Let $E = \bigcap E_i$ so $E_1 - E = (E_1 - E_2) \cup (E_2 - E_3) \cup \cdots$ and $\mu(E_1 - E) = \sum_{i=1}^{\infty} \mu(E_i - E_{i+1})$.

(ii) Show that $\mu(E_1) < \infty$ (or $\mu(E_n) < \infty$ for some n) is a necessary assumption.

Problem 9. Let $\{E_i\}$ be a sequence of measurable sets such that $E_1 \subset E_2 \subset E_3 \subset \cdots$. Show that $\mu\left(\bigcup E_i\right) = \lim \mu(E_i)$.

Problem 10. For any two sets E and F, define $E \triangle F$ by

$$E \triangle F = (E - F) \cup (F - E).$$

$E \triangle F$ is called the **symmetric difference** of E and F. Agree to identify sets E and F if $E \triangle F$ has measure zero. (Cf. Problem 5, Chapter 3.) Define a function d on pairs of subsets of $(0, 1)$ as follows:

$$d(E, F) = \mu(E \triangle F).$$

Show that d is a metric on (equivalence classes of) measurable sets. Notice that the triangle inequality - the only non-obvious metric property - implies that the relation $E \equiv F$, defined by $\mu(E \triangle F) = 0$, is an equivalence relation, thus providing the justification for identifying sets E and F if $E \equiv F$. Show that $\mu(E) = \mu(F)$ if $E \equiv F$, so that μ does not object to the identification of equivalent sets. Show that μ is continuous with respect to the metric d; *i.e.*, if $d(E_n, E) \longrightarrow 0$, then $\mu(E_n) \longrightarrow \mu(E)$. Is μ uniformly continuous? Is the restriction to subsets of $(0, 1)$ necessary? ◀

Problem 11. The operation Δ has some interesting properties which might appeal to those with an algebraic bent. For example, is Δ an associative operation? How does the operation Δ interact with $\cap$, $\cup$, $'$? Show that if intersection is interpreted as multiplication, and symmetric difference as addition, then the subsets of X (or the measurable subsets of X) form a commutative ring with identity. ◀

Problem 12. The Cantor Set. Each number in $[0, 1]$ can be written as a ternary series:

$$x = a_1/3 + a_2/3^2 + a_3/3^3 + \cdots,$$

where all a_i are $0, 1$, or 2. Some numbers have two such representations *e.g.*,

$$\begin{aligned} \frac{2}{3} &= 2/3 + 0/9 + 0/27 + \cdots \\ &= 1/3 + 2/9 + 2/27 + \cdots. \end{aligned}$$

Let U_1 be open middle third of $[0, 1]$; *i.e.*, $U_1 = (\frac{1}{3}, \frac{2}{3})$. Let U_2 be the two intervals which are the middle thirds of the two intervals in $[0, 1] - U_1$; *i.e.*, $U_2 = (\frac{1}{9}, \frac{2}{9}) \cup (\frac{7}{9}, \frac{8}{9})$. In general, let U_{n+1} be the union of all open middle thirds of the closed intervals in $[0,1] - \bigcup_{i=1}^{n} U_i$. The Cantor set is $[0,1] - \bigcup U_n$. Show:

(i) The Cantor set is a closed set of measure zero.

(ii) The Cantor set consists of exactly those points in $[0,1]$ which can be written with a ternary expansion with all $a_i = 0$ or 2. (For example, $\frac{1}{3}, \frac{2}{3}, \frac{1}{9}, \frac{2}{9}, \frac{7}{9}, \frac{8}{9}, \frac{1}{27}, \frac{2}{27}, \frac{7}{27}, \frac{8}{27}, \ldots$). Equivalently, show that $\bigcup U_n$ consists of those points whose ternary expansion must have some $a_i = 1$. Show that $\frac{1}{4}$ is in the Cantor set.

(iii) Show that all points of $[0,1]$ can be expressed as a binary expansion,

$$x = b_1/2 + b_2/2^2 + b_3/2^3 + \cdots,$$

where each b_i is 0 or 1.

(iv) Since both the Cantor set and $[0,1]$ can be put in a 1–1 correspondence with all sequences onto a two element set, the Cantor set is an uncountable set, and thus is an example of an uncountable set of measure zero.

(v) Show that each point of the Cantor set is the limit of a sequence of distinct points of the Cantor set.

(vi) Show how to define a closed nowhere dense subset of $[0,1]$ with arbitrary measure between 0 and 1 by modifying the above procedure. For example, to get a set of measure 1/2 we remove open intervals with total length 1/2 as follows: let U_1 be the open interval of length 1/4 centered in $[0,1]$. Then $[0,1] - U_1$ consists of two closed intervals whose lengths are less than 1/2. From these two closed intervals remove equal centered open intervals with lengths totalling $\frac{1}{8}$. Etc. ◀|||

Problem 13. Let E_1 and E_2 be disjoint measurable sets. Draw the appropriate figure similar to Fig. 1, showing E_1, E_2 and an arbitrary test set T. Label the subsets of T as follows: $E_1 \cap T = T_1$, $E_2 \cap T = T_2$, $(E_1 \cup E_2)' \cap T = T_3$. Write out the proof that $m(T_1 \cup T_2) + m(T_3) = m(T)$ which shows that $E_1 \cup E_2$ is measurable in this special case of Proposition 6. ◀|||

Problem 14. Here is the historical definition of measurable set. An open subset of $(0,1)$ is a countable union of disjoint open intervals. If $U = \bigcup (a_i, b_i)$, then define $m(U) = \sum (b_i - a_i)$. If F is a closed subset of $(0,1)$, and $U = (0,1) - F$, then define

$m(F) = 1 - m(U)$. Define outer measure m^* and inner measure m_* as follows:

$$m^*(E) = \inf\{m(U) : E \subset U,\ U \text{ open}\},$$
$$m_*(E) = \sup\{m(F) : F \subset E,\ F \text{ closed}\}.$$

A set E is measurable if and only if $m^*(E) = m_*(E)$. Show that $m_*(E) = 1 - m^*(E')$ so $m^*(E) = m_*(E)$ is the same as $m^*(E) + m^*(E') = 1$.

5 THE LEBESGUE INTEGRAL FOR BOUNDED FUNCTIONS

In this chapter we define the Lebesgue integral of a bounded function on a set of finite measure, and characterize the integrable functions. The development will closely parallel that for the Riemann integral to emphasize the similarity of the two ideas.

In the next chapter the definition is extended to unbounded functions and sets of infinite measure. The integral of a general non-negative function - possibly unbounded, and possibly defined on a set of infinite measure - will be the limit (supremum) of the integrals of smaller functions which are bounded and defined on sets of finite measure. We will call the bounded integrable functions on sets of finite measure the **primary functions**. Thus the integral of any non-negative f is defined to be the sup of the integrals of smaller primary functions. The integral of a non-positive function f is the negative of the integral of $-f$. The integral of a sometimes positive, sometimes negative function is obtained by considering the positive part and the negative part separately, and adding the results.

Now back to primary functions. Let S be a set of finite measure, by which we will always henceforth mean a **measurable** set of finite measure. A **partition** of S is a **finite** family $\{E_1, \ldots, E_n\}$ of disjoint measurable subsets of S whose union is S. In case S is an interval $[a, b]$ and P is a partition in the earlier sense, $P = \{x_0, x_1, \ldots, x_n\}$, we will now understand that P denotes the partition of $S = [x_0, x_n]$ into the disjoint sets $\{[x_0, x_1), [x_1, x_2), \ldots, [x_{n-1}, x_n]\}$.

If P and Q are partitions of S, then Q is a **refinement** of P, denoted $Q \succ P$ or $P \prec Q$, provided each $F \in Q$ is a subset of some $E \in P$. If $P = \{E_i\}$ and Q is a refinement of P, we will write $Q = \{F_{ij}\}$ to indicate that $E_i = \bigcup_j F_{ij}$ for each $E_i \in P$. Notice that if $P = \{E_i\}$ is a partition of S, then $\mu(S) = \sum \mu(E_i)$.

If f is a bounded function on a set S of finite measure, and $P = \{E_i\}$ is a partition of S, we define upper and lower sums for f and P exactly as in Chapter 1:

$$\begin{aligned} m_i &= \inf \{f(x) : x \in E_i\}, \\ M_i &= \sup \{f(x) : x \in E_i\}, \\ L(f,P) &= \sum_{i=1}^{n} m_i \, \mu(E_i), \\ U(f,P) &= \sum_{i=1}^{n} M_i \, \mu(E_i). \end{aligned}$$

If S is an interval and P is a partition of S into intervals, then $U(f,P)$ and $L(f,P)$ have exactly the same meaning as in Chapter 1.

Proposition 1. *If S is a set of finite measure and $m \le f(x) \le M$ for all $x \in S$, and P, Q are partitions of S with $Q \succ P$, then*

$$m\mu(S) \le L(f,P) \le L(f,Q) \le U(f,Q) \le U(f,P) \le M\mu(S).$$

Proof. Let $P = \{E_i\}$ and $Q = \{F_{ij}\}$ with $\bigcup_j F_{ij} = E_i$ for each i. Let

$$\begin{aligned} m_i &= \inf \{f(x) : x \in E_i\} \\ m_{ij} &= \inf \{f(x) : x \in F_{ij}\}. \end{aligned}$$

Then clearly $m_i \le m_{ij}$ for all i, j, so

$$\begin{aligned} L(f,P) &= \sum m_i \, \mu(E_i) \\ &= \sum_i m_i \sum_j \mu(F_{ij}) \\ &\le \sum_{i,j} m_{ij} \, \mu(F_{ij}) \\ &= L(f,Q). \end{aligned}$$

The proof that $U(f, Q) \le U(f, P)$ is similar, and the remaining inequalities are obvious. ▮

Problem 1. Show that every lower sum $L(f, P)$ is less than or equal to every upper sum $U(f, Q)$. ◂

If S is a set of finite measure and f is bounded on S, then f is **(Lebesgue) integrable** on S if and only if $\sup_P L(f, P) = \inf_Q U(f, Q)$. We write $\int_S f$ for the common value if f is integrable.

We observe that f is integrable on S if and only if there is for every $\varepsilon > 0$ a partition P such that $U(f, P) - L(f, P) < \varepsilon$.

Proposition 2. *If f is Riemann integrable on $[a, b]$, then f is Lebesgue integrable on $[a, b]$ and the integrals are the same.*

Proof. If f is Riemann integrable then there is a partition P of $[a, b]$ into intervals such that $U(f, P) - L(f, P) < \varepsilon$, so f is also Lebesgue integrable. The value of either integral lies between any lower sum and any upper sum, so the Riemann and Lebesgue integrals clearly coincide. ▮

For the Riemann integral we partitioned an interval $[a, b]$ into a finite number of subintervals:

$$[a, b] = [a, x_1) \cup [x_1, x_2) \cup \cdots \cup [x_{n-1}, b].$$

For the Lebesgue integral we partition an arbitrary measurable set S into a finite number of measurable subsets:

$$S = E_1 \cup E_2 \cup \cdots \cup E_n, \quad E_i \cap E_j = \emptyset \text{ if } i \ne j.$$

Since μ is countably additive, we might ask whether we should not consider instead partitions of S into countably many disjoint sets:

$$S = E_1 \cup E_2 \cup \cdots, \quad E_i \cap E_j = \emptyset \text{ if } i \ne j.$$

The problem below asks you to show that countable partitions with their corresponding lower and upper sums would give an equivalent definition of integrable function. Since the sup of lower sums over all countable partitions is larger than the sup over finite partitions, and upper sums similarly could be smaller if countable partitions were allowed, it is *a priori* easier for f to be integrable if countable partitions are allowed.

Problem 2. Show that the definition of integrability for a bounded function does not change if countable partitions are allowed.

The **characteristic function** of a set E, denoted χ_E, is the function which is one on the set and zero elsewhere. A **simple function** is a function φ defined on a set S of finite measure such that φ has a finite range $\{y_1, y_2, \ldots, y_n\}$ and for some partition $P = \{E_1, \ldots, E_n\}$ of S,

$$\varphi = \sum_{i=1}^{n} y_i \chi_{E_i} .$$

Problem 3. If φ is the simple function above, then φ is integrable and $\int_S \varphi = \sum_{i=1}^{n} y_i \mu(E_i)$.

The usual definition of "f is integrable over S" is

$$\sup\left\{\int \varphi : \varphi \le f\right\} = \inf\left\{\int \psi : \psi \ge f\right\}$$

where φ and ψ range over simple functions defined on S. This is of course just a different method for expressing the same idea as our definition.

If f is continuous on an interval $[a, b]$ then f is Riemann integrable on $[a, b]$. The proof consists in showing that since f is uniformly continuous, each $M_i - m_i$ will be less than any given $\varepsilon > 0$ provided P is any sufficiently fine partition of $[a, b]$ into intervals. This implies

$$U(f, P) - L(f, P) = \sum (M_i - m_i)\,\Delta x_i < \varepsilon(b - a) ,$$

so f is Riemann integrable. A bounded function f will be Lebesgue integrable on a set S of finite measure, by the same argument, if there is a partition $P = \{E_i\}$ of S so that $M_i - m_i < \varepsilon$ for each

i. There will obviously be such a partition provided each set of the form $\{x \in S : a \le f(x) < a + \varepsilon\}$ is measurable; specifically, we can let

$$E_i = \{x \in S : m + i\varepsilon \le f(x) < m + (i+1)\varepsilon\} \ , i = 0, 1, 2, \ldots,$$

where m is a lower bound for f on S. These sets are disjoint, and a finite number of them will form a partition of S, since f is bounded, provided only that each of these sets is measurable. Accordingly, we agree that f **is measurable on** S provided $\{x : a \le f(x) < b\}$ is measurable for all a, b. Notice that if f is measurable on S, then S is necessarily a measurable set. The definition of measurable function applies to all functions f, bounded or not, and all measurable sets S, whether or not they have finite measure. In this chapter we are only concerned with the integrals of bounded functions on sets of finite measure, but in Chapter 7 we will consider unbounded measurable functions on sets with infinite measure.

Problem 4. If f is measurable on S and $g = f$ except on a zero measure subset of S, then g is measurable on S. ◀

Problem 5. Every simple function is measurable. ◀

Problem 6. If f is continuous on $[a, b]$, then f is measurable on $[a, b]$. Hint: Show that $\{x \in [a, b] : f(x) \ge \alpha\}$ is a closed set (and therefore measurable) for each α. Then use

$$\{x \in [a, b] : \alpha \le f(x) < \beta\} =$$
$$\{x \in [a, b] : f(x) \ge \alpha\} - \{x \in [a, b] : f(x) \ge \beta\} \ .$$ ◀

Problem 7. If f is measurable then $|f|$ is measurable. ◀

Proposition 3. *If f is a bounded measurable function on a set S of finite measure, then f is Lebesgue integrable on S.*

Proof. Let $-M \le f(x) < M$ for all $x \in S$. Let N be a large integer, and let

$$E_i = \{x : -M + (i-1)/N \le f(x) < -M + i/N\}$$

for $i = 1, 2, \ldots, 2MN$. Then $P = \{E_i\}$ is a partition of S and

$$U(f,P) - L(f,P) \le \sum \frac{1}{N}\mu(E_i) = \mu(S)/N\,.$$

The converse of Proposition 3 is also true for bounded functions (Proposition 6 below), so that measurability is equivalent to integrability for bounded functions on sets of finite measure. The great virtue of this characterization is not the fact that more functions are Lebesgue integrable, but the fact that pointwise limits of measurable functions are measurable, as we show below. Hence if $f = \lim f_n$ with each f_n integrable, then $\int f$ makes sense provided only that f is bounded. It is this kind of result which makes it very much easier to deal with limits of integrals and integrals of limits in Lebesgue integration.

Problem 8. Show that there is a sequence $\{f_n\}$ of Riemann integrable functions on $[0,1]$, all with the same integral, such that $f_n(x) \longrightarrow f(x)$ for all $x \in [0,1]$ and f is not Riemann integrable.

Now we proceed to show that every integrable function is measurable. We show that an integrable function is the pointwise limit of simple functions, which are necessarily measurable, and that every pointwise limit of measurable functions is measurable. We first introduce some useful alternative criteria for measurability.

Proposition 4. *Each of the following conditions is necessary and sufficient for f to be measurable:*
(i) $\{x : f(x) \ge a\}$ is measurable for all a;
(ii) $\{x : f(x) < a\}$ is measurable for all a;
(iii) $\{x : f(x) > a\}$ is measurable for all a;
(iv) $\{x : f(x) \le a\}$ is measurable for all a;
(v) $\{x : a < f(x) < b\}$ is measurable for all a, b.

Proof. The sets in (i) and (ii) are complements, and similarly for the sets in (iii) and (iv). If f is measurable, then $\{x : f(x) \ge a\}$ is the countable union of the measurable sets $\{x : a \le f(x) <$

$a+n\}$, $n = 1, 2, \ldots$. Conversely, if $\{x : a \le f(x)\}$ is measurable for all a, then

$$\{x : a \le f(x) < b\} = \{x : a \le f(x)\} - \{x : b \le f(x)\}$$

is measurable for all a, b, and hence f is measurable. The other equivalencies are proved similarly, using the fact that measurable sets are closed under countable unions and intersections, and under complementation. ▮

Problem 9. Complete the proof of Proposition 4. ▮

In this chapter we consider the integral only for bounded functions. In later chapters, however, we will consider unbounded functions, and indeed functions that take the values $+\infty$ or $-\infty$, since these values can arise as limits of sequences of integrable functions. Accordingly, we agree that such an extended real valued function f is measurable provided $\{x : a \le f(x) < b\}$ is measurable for all a, b, and the sets $\{x : f(x) = +\infty\}$ and $\{x : f(x) = -\infty\}$ are both measurable.

Proposition 5. *If $\{f_n\}$ is a sequence of measurable functions on a measurable set S, then* $\sup f_n$, $\inf f_n$, $\limsup f_n$, *and* $\liminf f_n$ *are measurable functions. If* $\lim f_n(x)$ *exists for all x, then the limit is a measurable function.*

Proof. To show $\sup f_n$ is measurable we verify condition (iii) of Proposition 4. Since

$$\{x : \sup f_n(x) > a\} = \bigcup_n \{x : f_n(x) > a\},$$

$\{x : \sup f_n(x) > a\}$ is a countable union of measurable sets if each f_n is measurable. Similarly,

$$\{x : \inf f_n(x) < a\} = \bigcup_n \{x : f_n(x) < a\},$$

so $\inf f_n$ is measurable.

If $\sup f_n$ takes the value $+\infty$, then

$$\{x : \sup f_n(x) = +\infty\} = \bigcap_N \bigcup_n \{x : f_n(x) > N\},$$

and this set is measurable. A similar equality holds for the set where $\inf f_n(x) = -\infty$.

Since

$$\limsup f_n(x) = \inf_n \sup_{k\geq n} f_k(x)\,,$$
$$\liminf f_n(x) = \sup_n \inf_{k\geq n} f_k(x)\,,$$

both $\limsup f_n$ and $\liminf f_n$ are measurable. If $\lim f_n(x)$ exists for all x, then $\lim f_n = \limsup f_n = \liminf f_n$. ▮

We will use the phrase **almost everywhere**, abbreviated **a.e.**, to mean "except on a set of measure zero." Hence "$f = g$ a.e." means that $\{x : f(x) \neq g(x)\}$ has measure zero, and "$f \geq n$ a.e." means $\{x : f(x) < n\}$ has measure zero.

Proposition 6. *If f is a bounded function which is integrable on a set S of finite measure, then f is the a.e. pointwise limit of simple functions, and hence f is measurable.*

Proof. For each n we let P_n be a partition of S such that $U(f,P_n) - L(f,P_n) < 1/n$. We can assume that $P_1 \prec P_2 \prec P_3 \prec \cdots$ by replacing each P_n by the common refinement of its predecessors. Let $P_n = \{E_{ni}\}$ and

$$m_{ni} = \inf\{f(x) : x \in E_{ni}\}$$
$$M_{ni} = \sup\{f(x) : x \in E_{ni}\}\,.$$

Let φ_n be the simple function which has the value m_{ni} on E_{ni}, and let ψ_n be the function which is M_{ni} on E_{ni}. Then

$$\varphi_n \leq f \leq \psi_n\,;$$
$$L(\varphi_n, P_n) = L(f, P_n); \quad U(\psi_n, P_n) = U(f, P_n)\,.$$

The functions φ_n increase to some measurable function $g \leq f$, and the ψ_n decrease to some measurable $h \geq f$. We will show that $g = h$ a.e., so that $g = f = h$ a.e. and f is measurable.

Suppose on the contrary that $h - g > 0$ on a set A of positive measure. Then (Problem 10 below) there is $p > 0$ so $h - g > p$ on a set A_0 of positive measure. For each n let P_n^* be the refinement of P_n gotten by replacing each $E_{ni} \in P_n$ by the two sets $E_{ni} \cap A_0$ and $E_{ni} \cap A_0'$. Then on any set $E_{ni} \cap A_0$ we have

$$\psi_n - \varphi_n \geq h - g > p\,.$$

Since $\psi_n = M_{ni}$ and $\varphi_n = m_{ni}$ on $E_{ni} \cap A$ it follows that

$$\begin{aligned} U(f,P_n^*) - L(f,P_n^*) &\geq p\sum_i \mu(E_{ni} \cap A_0) \\ &= p\mu(A_0)\,. \end{aligned}$$

Since $P_n^* \succ P_n$, we also have for all n,

$$U(f,P_n^*) - L(f,P_n^*) \leq U(f,P_n) - L(f,P_n) < \frac{1}{n}\,,$$

which contradicts the preceding estimate, so $h = f = g$ a.e. ▮

Problem 10. Show that if k is a measurable function and $\{x : k(x) > 0\}$ has positive measure, then $\{x : k(x) \geq p\}$ has positive measure for some $p > 0$. ◀

Problem 11. If f is a bounded function which is measurable on a set S of finite measure, and T is a measurable subset of S, then f is integrable over T. ◀

Problem 12. Show that if f is integrable over S, then $\int_S f = \lim_P L(f,P) = \lim_P U(f,P)$ where the partitions are ordered by refinement. ◀

Problem 13. If f is bounded on $[a,b]$ and continuous except at a finite number of points, then f is measurable and hence integrable. ◀

6 PROPERTIES OF THE INTEGRAL

We will prove the linearity properties of the integral by showing that the integral is a limit of Riemann sums. First we need to know that kf and $f + g$ are measurable if f and g are.

Proposition 1. *If f and g are measurable on S, then kf is measurable for every constant k, and $f + g$ is measurable.*

Proof. It is clear that kf is measurable if f is, so we consider $f + g$. The inequality $f(x) + g(x) > a$ is equivalent to $f(x) > a - g(x)$, which holds if and only if there is a rational number r such that

$$f(x) > r \quad \text{and} \quad r > a - g(x).$$

Hence

$$\{x : f(x) + g(x) > a\} = \bigcup_r \{x : f(x) > r\} \cap \{x : g(x) > a - r\},$$

where the union is over all rationals r. The right side is a countable union of measurable sets. ▮

If f and g are bounded measurable functions on a set S of finite measure, then kf and $f + g$ are integrable over S. Now we write $\int f$ as a limit of Riemann sums. If S is a set of finite measure and $P = \{E_i\}$ is a partition of S, then a **choice function** for P is a finite sequence $\{c_i\}$ with $c_i \in E_i$ for each i. The **Riemann sum** for f, P, c is the usual sum

$$R(f, P, c) = \sum_i f(c_i)\mu(E_i).$$

The partitions P of S form a directed set under the partial ordering of refinement, and the pairs (P, c) are ordered according to the ordering on P; *i.e.*, $(P,c) \succ (P',c')$ means $P \succ P'$ (P is a refinement of P'). The Riemann sums $R(f,P,c)$ are a net on the partially ordered pairs (P,c). In this context the limit condition for nets reads as follows: $R(f,P,c) \longrightarrow I$, or $\lim_P R(f,P,c) = I$, if and only if for each $\varepsilon > 0$ there is a partition P_0 such that $|R(f,P,c) - I| < \varepsilon$ whenever $P \succ P_0$, and c is a choice for P. Although $R(f,P,c)$, for fixed f, is a function of the pair (P,c), the pairs are ordered only in terms of P, so we write $\lim_P R(f,P,c)$ instead of the correct but more cumbersome $\lim_{(P,c)} R(f,P,c)$.

Proposition 2. *If f is a bounded function which is integrable on the finite measure set S, then $R(f,P,c) \longrightarrow \int_S f$.*

Proof. If P is any partition of S, and c is any choice function for P, then

$$L(f,P) \le R(f,P,c) \le U(f,P). \tag{1}$$

If f is integrable then for any $\varepsilon > 0$ there is a partition P_0 so that $U(f,P) - L(f,P) < \varepsilon$ for all $P \succ P_0$. Hence

$$\left| R(f,P,c) - \int_S f \right| < \varepsilon$$

for all $P \succ P_0$ and all c; *i.e.*, $R(f,P,c) \longrightarrow \int_S f$. ▮

Notice that if $\{U(f,P)\}$, $\{L(f,P)\}$, $\{R(f,P,c)\}$ were nets on the same directed set, the consequence $R(f,P,c) \longrightarrow \int_S f$ would follow immediately from the inequality (1) and the fact that $\lim_P L(f,P) = \lim_P U(f,P) = \int_S f$. As it is, the directed set for the Riemann sums is the much larger directed set consisting of all pairs (P,c) instead of just all partitions P.

The following proposition shows that the net of Riemann sums $\{R(f,P,c)\}$ can not distinguish between f and a function which equals f almost everywhere.

Proposition 3. *If f and g are arbitrary functions on a finite measure set S, and $f = g$ a.e., then*

$$\lim_P R(f,P,c) = \lim_P R(g,P,c);$$

i.e., one limit exists if and only if the other does, and then the limits are equal. The functions f and g are not assumed to be bounded or measurable.

Proof. Let $f = g$ except on $A \subset S$, where $\mu(A) = 0$. Assume $\lim_P R(f,P,c) = L$. Let $\varepsilon > 0$ and choose a partition P_0 so that $|R(f,P,c) - L| < \varepsilon$ if $P \succ P_0$. Let P_1 be the refinement of P_0 obtained by replacing each set E_i of P_0 by the two sets $E_i \cap A$ and $E_i \cap A'$. Many of these sets may be empty, but that doesn't matter. Let $P \succ P_1$. If $P = \{F_i\}$, then each F_i is a subset of some $E_j \cap A$, so $F_i \subset A$ and $\mu(F_i) = 0$, or F_i is a subset of some $E_j \cap A'$, so $f = g$ on F_i. Therefore, if $P \succ P_1 \succ P_0$,

$$\begin{aligned} R(f,P,c) &= \sum f(c_i)\mu(F_i) \\ &= \sum g(c_i)\mu(F_i) = R(g,P,c)\,. \end{aligned}$$

Since $|R(f,P,c)-L| < \varepsilon$ if $P \succ P_0$, it follows that $|R(g,P,c)-L| < \varepsilon$ if $P \succ P_1$; *i.e.*, if $\lim R(f,P,c)$ exists, then so does $\lim R(g,P,c)$ and the limits are equal. The situation is symmetric in f and g, so we are done. ||||||

Notice from Proposition 3 that Riemann sums can converge for an unbounded function, which is unlike the situation for the Riemann integral. For example, let g be a bounded measurable function on S, so $R(g,P,c) \longrightarrow \int_S g$. Let $f = g$ except on some countable set $\{x_n\}$, and let $f(x_n) = n$, so f is unbounded, but $f = g$ a.e. By Proposition 3, $\lim R(f,P,c) = \lim R(g,P,c) = \int_S g$. The next two problems point out that the only way Riemann sums can converge is for the function to be essentially equal to an integrable function.

Problem 1. If g is a bounded function on a set S of finite measure, and $R(g,P,c) \longrightarrow I$, then g is integrable (hence measurable) and $\int_S g = I$. Hint: Cf. Proposition 3, Chapter 2. ||||

Problem 2. If f is any function on a set S of finite measure, and $R(f,P,c) \longrightarrow I$, then there is a bounded function g such that $f = g$ a.e. and $R(g,P,c) \longrightarrow I$. Hence $R(f,P,c) \longrightarrow I$ if and only if $f = g$ a.e. for some bounded integrable function g. ||||

The following proposition is now a simple consequence of the fact that the integral is a *bona fide* limit.

Proposition 4. *If f and g are bounded measurable functions on a set S of finite measure, and k is a constant, then*

(i) $\int_S kf = k\int_S f$

(ii) $\int_S (f+g) = \int_S f + \int_S g$

(iii) $|\int_S f| \le \int_S |f|$

Proof. (i)

$$\begin{aligned}\int_S kf &= \lim_P R(kf,P,c)\\ &= \lim_P \sum kf(c_i)\mu(E_i)\\ &= \lim_P k\sum f(c_i)\mu(E_i)\\ &= \lim_P kR(f,P,c)\\ &= k\lim_P R(f,P,c)\\ &= k\int_S f.\end{aligned}$$

Problem 3. (i) Verify that the net $\{R(f+g,P,c)\}$ is the sum of the nets $\{R(f,P,c)\}$ and $\{R(g,P,c)\}$ and use this to prove part (ii) of Proposition 4.

(ii) Verify that $R(|f|,P,c) \ge |R(f,P,c)|$ for all (P,c), and use this to prove part (iii) of Proposition 4. Why does the net $\{R(|f|,P,c)\}$ converge?

Proposition 5. *If f is a bounded measurable function on a finite measure set S, and T is a measurable subset of S, then f is integrable over T, and*

$$\int_T f = \int_S f\chi_T.$$

Proof. Let $\mathcal{Q}$ be all partitions Q of S such that every $E \in Q$ is either a subset of T or disjoint from T. Every partition P of S can be refined to get such a partition $Q \in \mathcal{Q}$, so all integrals over S can be expressed as limits of sums $R(f,Q,c)$ as Q ranges over

$\mathcal{Q}$. Hence if $Q \in \mathcal{Q}$ and $Q = \{E_i \cap T, E_i \cap T'\}$, then

$$\begin{aligned} \int_S f\chi_T &= \lim_{Q \in \mathcal{Q}} R(f\chi_T, Q, c) \\ &= \lim_Q \sum_i f(c_i)\mu(E_i \cap T). \qquad (2) \\ &= \int_T f. \end{aligned}$$

The last equality holds because every Riemann sum for $\int_T f$ is one of the sums in (2). ∎

Corollary. *If f is a bounded measurable function on $A \cup B$, where A and B are disjoint measurable sets, then*

$$\int_{A \cup B} f = \int_A f + \int_B f.$$

Proof. $f = f \cdot \chi_A + f \cdot \chi_B$. ∎

Most of the functions one wants to integrate are continuous – perhaps even analytic. For such functions there is no difference between the Riemann and Lebesgue integrals over a bounded closed interval. The Lebesgue integral, however, is much more accommodating in the matter of limit theorems. For the Riemann integral one must generally know that $f_n \longrightarrow f$ uniformly to conclude that $\int f_n \longrightarrow \int f$. For the Lebesgue integral, pointwise convergence is enough provided the functions stay uniformly bounded. The reason for this is that pointwise convergence is nearly uniform on finite measure sets. We make this idea precise in the next two propositions.

Proposition 6. *If $\{f_n\}$ is a sequence of measurable functions on a finite measure set S, and $f_n \longrightarrow f$ pointwise on S, then given $\varepsilon > 0$ and $\delta > 0$, there is a measurable set E of measure less than δ and a number N so that $|f_k(x) - f(x)| < \varepsilon$ for all $k \geq N$ and all $x \in S - E$.*

Proof. Let

$$F_n = \{x \in S : |f_k(x) - f(x)| \geq \varepsilon \text{ for some } k \geq n\}.$$

The sets F_n are measurable, and decreasing, and $\bigcap F_n = \emptyset$ because $f_n(x) \longrightarrow f(x)$ for all x. Since $\mu(F_1) < \infty$, $\lim \mu(F_n) = 0$. Let $\mu(F_N) < \delta$. For $x \in S - F_N$, $|f_k(x) - f(x)| < \varepsilon$ for all $k \geq N$. ▮

Problem 4. If $f_n \longrightarrow f$ a.e. the same result holds. ◀

The next proposition gives the form in which it is easiest to remember and apply the above result.

Proposition 7. *(Egoroff's Theorem) If $\{f_n\}$ is a sequence of measurable functions on a finite measure set S, and $f_n \longrightarrow f$ pointwise on S, then for every $\delta > 0$ there is a measurable set $E \subset S$ of measure less than δ so that $f_n \longrightarrow f$ uniformly on $S - E$.*

Proof. For each n we find a set E_n of measure less than $\delta/2^n$ and a number N_n so that $|f_k(x) - f(x)| < \frac{1}{n}$ for $k \geq N_n$ and $x \in S - E_n$. Let $E = \bigcup E_n$, so that $\mu(E) < \delta$. If $x \in S - E$ then given $\varepsilon > 0$ there is N (any N_n with $\frac{1}{n} < \varepsilon$) so that $|f_k(x) - f(x)| < \varepsilon$ if $k \geq N$. ▮

For uniformly bounded sequences, the limit of the integrals is the integral of the limit. This follows directly from Egoroff's Theorem, as we show next.

Proposition 8. *(Bounded Convergence Theorem). If $\{f_n\}$ is a sequence of measurable functions on a finite measure set S, and the functions f_n are uniformly bounded on S, and $f_n(x) \longrightarrow f(x)$ pointwise on S, then*

$$\lim \int_S f_n = \int_S \lim f_n = \int_S f.$$

Proof. Let $\varepsilon > 0$. Let E be a measurable set of measure less than ε such that $f_n \longrightarrow f$ uniformly off E.

Let $|f_n(x)| \leq M$ for all n and all $x \in S$. Then

$$\begin{aligned}\left|\int_S (f_n - f)\right| &\leq \int_S |f_n - f| \\ &= \int_{S-E} |f_n - f| + \int_E |f_n - f| \\ &< \int_{S-E} |f_n - f| + 2M\varepsilon .\end{aligned}$$

Since $f_n \longrightarrow f$ uniformly on $S - E$, there is N so that $|f_n - f| < \varepsilon$ on $S - E$ if $n \geq N$. Hence if $n \geq N$,

$$\begin{aligned} \left| \int_S (f_n - f) \right| &< \varepsilon\mu(S - E) + 2M\varepsilon \\ &= \varepsilon[\mu(S - E) + 2M] \\ &\leq \varepsilon\left[\mu(S) + 2M\right]. \end{aligned}$$

Since ε is arbitrary and $\mu(S) < \infty$, the result follows. ▮

The hypotheses of the Bounded Convergence Theorem require that $f_n \longrightarrow f$ pointwise with all f_n remaining in some fixed finite area. Specifically, we require that all f_n lie in the rectangle $S \times [-M, M]$, where $\mu(S) < \infty$. If the functions are allowed to wander outside a fixed finite area the result can fail as the following problem shows.

Problem 5. Let $S = (0, 1)$. Give an example of bounded measurable functions f_n on S so that $\int_S f_n = 1$ for all n and $f_n(x) \longrightarrow 0$ for all $x \in S$. ◀

Problem 6. (i) If f is a measurable non-negative function on $[0, 1]$ and $\int f d\mu = 0$, then $f = 0$ a.e.

(ii) If f and g are bounded measurable functions on a set S of finite measure, and $f \leq g$ a.e., then $\int_S f \leq \int_S g$. ◀

Problem 7. If $0 \leq f_n \leq h \leq g_n \leq M$ on $[a, b]$ for all n, where $\{f_n\}$, $\{g_n\}$ are respectively increasing and decreasing sequences of measurable functions with $\lim \int f_n = \lim \int g_n$, then h is measurable and $\int h = \lim \int f_n$. ◀

Problem 8. Show that almost everywhere convergence is sufficient in the Bounded Convergence Theorem. ◀

Problem 9. Let $f_n(x) = nx/(1 + n^2x^2)$ for $0 \leq x \leq 1$. (i) Show that $f_n(x) \longrightarrow 0$ for all $x \in [0, 1]$, but the convergence is not uniform.

(ii) Does the Bounded Convergence Theorem apply? ◀

Problem 10. If f and g are measurable functions, then fg is measurable. Hint: $fg = \frac{1}{4}[(f + g)^2 - (f - g)^2]$, so it suffices to show that h^2 is measurable if h is measurable. ◀

Problem 11. If f is bounded on $[a,b]$, then f is Riemann integrable on $[a,b]$ if and only if f is continuous almost everywhere; *i.e.*, the set where f is discontinuous has measure zero. Verify the following arguments. Let $\{P_n\}$ be a sequence of partitions of $[a,b]$ such that $\|P_n\| < \frac{1}{n}$ and $P_1 \prec P_2 \prec P_3 \prec \cdots$. Let I_{ni} be the ith subinterval of P_n. Let M_{ni}, m_{ni} be the sup and inf of f on I_{ni}. Let g_n be the step function which is m_{ni} on I_{ni}, and let h_n be M_{ni} on I_{ni}. Then g_n and h_n are measurable, and $g_n \le f \le h_n$ for all n. The sequence $\{g_n\}$ increases to a measurable function $g \le f$, and $\{h_n\}$ decreases to a measurable function $h \ge f$. Show that if f is continuous at x_0, then $g(x_0) = h(x_0)$. Show conversely that if $h(x_0) - g(x_0) > \varepsilon$, then $M_{ni} - m_{ni} > \varepsilon$ whenever $x_0 \in I_{ni}$, and consequently there are points $u_n, v_n \in I_{ni}$ so $f(u_n) - f(v_n) > \varepsilon$. Since $\|P_n\| \longrightarrow 0$, $u_n \longrightarrow x_0$ and $v_n \longrightarrow x_0$, and f is discontinuous at x_0.

Now we know that f is continuous at x_0 if and only if $g(x_0) = h(x_0)$. Use the fact that $\int g_n = L(f, P_n)$, $\int h_n = U(f, P_n)$ to show that f is Riemann integrable if and only if the Lebesgue integral $\int (h - g) = 0$. By Problem 6 we know $\int (h - g) = 0$ if and only if $h = g$ a.e., so f is Riemann integrable if and only if f is continuous a.e.

Problem 12. Let C be the Cantor set (Problem 11, Chapter 4) and let D be a Cantor-like set of positive measure (*i.e.*, a nowhere dense closed subset of $[0,1]$ of positive measure). Are the characteristic functions χ_C and χ_D Riemann integrable?

7 THE INTEGRAL OF UNBOUNDED FUNCTIONS

In this chapter we define the integral $\int_S f$ in the cases where f is unbounded or S has infinite measure. These are situations where the Riemann integral would be called improper. For the Lebesgue integral the extension to unbounded functions and sets is more natural, and we will not need to stigmatize that situation with the "improper" terminology. We will now also consider extended real valued functions, which may take the values $+\infty$ or $-\infty$. Such functions f will arise naturally as limits of sequences, and indeed as limits of sequences $\{f_n\}$ such that $\int f_n$ converges to $\int f$.

The general definition of $\int_S f$ will coincide with the geometric idea of the net area under the graph; *i.e.*, the area above the x-axis minus the area below the x-axis. Both these areas will be required to be finite, in contradistinction to the improper Riemann integral. For example, the function which is $(-1)^n \frac{1}{n}$ on the interval $(n, n+1)$ is improperly Riemann integrable over $[1, \infty)$ since

$$-1 + \frac{1}{2} - \frac{1}{3} + \frac{1}{4} - \cdots$$

converges, and consequently

$$\lim_{x \to \infty} \int_1^x f$$

converges. This function is not Lebesgue integrable, since the positive area is

$$\frac{1}{2} + \frac{1}{4} + \frac{1}{6} + \cdots,$$

which is infinite. A Lebesgue integrable function is thus always absolutely integrable in the sense that if f is measurable, then f

is integrable if and only if $|f|$ is integrable. For these reasons we define the integral first for measurable **non-negative** functions. We then write any measurable function f as $f = f^+ - f^-$ where f^+ and f^- are non-negative:

$$f^+(x) = \max\{f(x), 0\} = f(x) \vee 0,$$
$$f^-(x) = \max\{-f(x), 0\} = (-f(x)) \vee 0\,.$$

The final definition will be

$$\int f = \int f^+ - \int f^-$$

provided both $\int f^+$ and $\int f^-$ are finite.

Problem 1. Show that $\lim\limits_{N\to\infty} \int_1^N \frac{\sin x}{x}\,dx$ exists (as a finite number), but $\lim\limits_{N\to\infty} \int_1^N \left|\frac{\sin x}{x}\right| dx$ does not.

Problem 2. Is $\frac{1}{x}\sin\frac{1}{x}$ improperly Riemann integrable on $[0,1]$? Is $\left|\frac{1}{x}\sin\frac{1}{x}\right|$? (cf. Problem 1).

Recall that the **primary functions** are the functions for which the integral has already been defined. That is, the primary functions are the bounded measurable functions defined on measurable sets of finite measure, or perhaps only defined a.e. on sets of finite measure. These functions f are all integrable; *i.e.*, $\int_S f$ makes sense as the finite limit of the nets of upper sums, lower sums, and Riemann sums.

Now let f be a non-negative, measurable, extended real-valued function defined on a measurable set S which might be all of $\mathbb{R}$. We define

$$\int_S f = \sup\left\{\int_S g : 0 \le g \le f, \quad g \text{ primary}\right\}.$$

Clearly this definition agrees with our earlier definition if f is a non-negative primary function. The integral makes sense for any non-negative f, but $\int_S f = \infty$ is a possibility. We say f is **integrable over** S if $\int_S f < \infty$.

Problem 3. If f is a non-negative measurable function on $\mathbb{R}$, let $f_M = f \wedge M$ on $[-M, M]$ and $f_M = 0$ outside $[-M, M]$. Show that $\lim\limits_{M\to\infty} \int_{\mathbb{R}} f_M = \int_{\mathbb{R}} f$.

Proposition 1. *If f and g are non-negative measurable (but not necessarily integrable) functions on S, and $k \geq 0$, then*

(i) $\int_S kf = k\int_S f$,

(ii) $\int_S (f+g) = \int_S f + \int_S g$.

(iii) If $f \leq g$, *then* $\int_S f \leq \int_S g$.

(iv) If $0 \leq f \leq g$ *and* g *is integrable, then* f *and* $g - f$ *are integrable, and* $\int_S g = \int_S f + \int_S (g-f)$.

Proof. Parts (i) and (iii) are immediate from the definition. To prove (ii), let h_1, h_2 be primary functions with $0 \leq h_1 \leq f$ and $0 \leq h_2 \leq g$. Then $h_1 + h_2$ is a primary function and $0 \leq h_1 + h_2 \leq f + g$. Consequently

$$\int_S (f+g) \geq \int_S (h_1 + h_2) = \int_S h_1 + \int_S h_2 .$$

Taking the sup over all such h_1 and h_2 we get

$$\int_S (f+g) \geq \int_S f + \int_S g .$$

To show the opposite inequality, let h be a primary function with $0 \leq h \leq f + g$. Let $h_1 = f \wedge h$ and $h_2 = h - h_1$. Then h_1 and h_2 are primary, and

$$0 \leq h_1 \leq f \text{ and } 0 \leq h_2 \leq g ,$$

so

$$\int_S h = \int_S (h_1 + h_2) = \int_S h_1 + \int_S h_2 \leq \int_S f + \int_S g .$$

Taking the sup over all primary $h \leq f + g$ we get

$$\int_S (f+g) \leq \int_S f + \int_S g . \quad ▮$$

Problem 4. Show that (iv) follows from (ii) and (iii) in Proposition 1. ◂

Corollary. *If f and g are integrable, then kf and $f+g$ are integrable.*

The next proposition is a basic fact - perhaps **the** basic fact - about convergence of integrals. If $f_n \geq 0$ and $f_n \longrightarrow f$ pointwise, then the areas under the f_n approach or exceed the area under f. That is

$$\liminf \int f_n \geq \int f \, .$$

(We will omit the domain of the integral when it is not material to the discussion.) We call attention to the examples

$$\begin{aligned} f_n &= \begin{cases} n \text{ on } \left(0, \dfrac{1}{n}\right) \\ 0 \text{ off } \left(0, \dfrac{1}{n}\right), \end{cases} \\ g_n &= \begin{cases} 1 \text{ on } (n, n+1) \\ 0 \text{ off } (n, n+1)\, . \end{cases} \end{aligned}$$

For both examples the functions approach zero pointwise, but $\int f_n = \int g_n = 1$ for all n. It is easy to see how to modify the examples so that, for example, $\int f_n = \int g_n = n$ for all n, with $\lim f_n = \lim g_n = 0$. If $f_n \longrightarrow f$ then the areas under the f_n will eventually soak up all the area under f, but the areas under the f_n may be too big.

Proposition 2. *(Fatou's Lemma) If $\{f_n\}$ is a sequence of non-negative measurable functions, and $f_n \longrightarrow f$ pointwise on $\mathbb{R}$, then*

$$\liminf \int f_n \geq \int f \, .$$

Proof. Let h be a primary function with $0 \leq h \leq f$. Let $h_n = f_n \wedge h$, so the functions h_n are primary and are uniformly bounded by a bound for h. Since $f_n \longrightarrow f \geq h$, $h_n \longrightarrow h$. By the Bounded Convergence Theorem, $\int h_n \longrightarrow \int h$. Since $\int f_n \geq \int h_n$ for all n

$$\liminf \int f_n \geq \lim \int h_n = \int h \, .$$

This last inequality holds for all primary $h \le f$, so

$$\liminf \int f_n \ge \int f . \quad ▮$$

Problem 5. (a) (Monotone Convergence Theorem). If $\{f_n\}$ is an increasing sequence of non-negative measurable functions, and $f_n \longrightarrow f$, then $\int f_n \longrightarrow \int f$.

(b) If g_n is a sequence of non-negative measurable functions, then

$$\int \sum g_n = \sum \int g_n . \quad ◀$$

Problem 6. If $\{f_n\}$ is any sequence of non-negative measurable functions,

$$\liminf \int f_n \ge \int \liminf f_n .$$

Hint: Let $g_n = \inf_{k \ge n} f_k$ so that g_n increases to $\liminf f_n$. ◀

The essential content of Fatou's Lemma - that $\liminf \int f_n$ can be too large only if the f_n are allowed an infinite area to play in - is summed up in the following basic theorem.

Proposition 3. *(Lebesgue Dominated Convergence Theorem - non-negative version) Let $\{f_n\}$ be a sequence of non-negative measurable functions such that $f_n \longrightarrow f$. If there is an integrable function g such that $0 \le f_n \le g$ for all n, then*

$$\int f_n \longrightarrow \int f .$$

Proof. Clearly f is integrable since $f \le g$. By Fatou's Lemma it is sufficient to show that

$$\limsup \int f_n \le \int f .$$

Since

$$0 \le g - f_n \longrightarrow g - f ,$$

Fatou's Lemma gives

$$\liminf \int (g - f_n) \geq \int (g - f) = \int g - \int f .$$

However,

$$\liminf \left(\int g - \int f_n \right) = \int g - \limsup \int f_n ,$$

so, since $\int g$ is finite,

$$\int g - \limsup \int f_n \geq \int g - \int f ,$$

$$\limsup \int f_n \leq \int f . \quad \blacksquare$$

Problem 7. If f is a non–negative measurable function on $\mathbb{R}$ and $\{E_i\}$ is a sequence of disjoint measurable sets with $\bigcup E_i = E$, then

$$\int_E f = \sum \int_{E_i} f . \quad \blacksquare$$

Now we complete the definition of the integral to include not necessarily positive functions. Let $f^+ = f \vee 0$ and $f^- = (-f) \vee 0$ so that f^+ and f^- are non–negative, and measurable if f is measurable. Since $f = f^+ - f^-$ we make the natural definition

$$\int f = \int f^+ - \int f^- .$$

We say f is **integrable** (or integrable over a set E) provided both f^+ and f^- are integrable (over E).

Problem 8. If f is integrable over $\mathbb{R}$, then f is integrable over any measurable set E. $\blacksquare$

The next proposition verifies the basic properties of the completed integral.

Proposition 4. *If f and g are integrable (over any measurable set S) and k is a constant, then kf and $f+g$ are integrable (over S), and*

(i) $\int kf = k\int f$;

(ii) $\int(f+g) = \int f + \int g$.

(iii) If $f \le g$, then $\int f \le \int g$.

Proof. (i) If $k > 0$, then $(kf)^+ = kf^+$ and $(kf)^- = kf^-$. Hence, if $k > 0$,

$$\begin{aligned}\int kf &= \int (kf)^+ - \int (kf)^- \\ &= k\int f^+ - k\int f^- = k\int f.\end{aligned}$$

The case $k < 0$ is left as an exercise.

(ii) Notice that $(f+g)^+$ is not generally equal to $f^+ + g^+$, so part (ii) is not automatic. However, if h_1 and h_2 are any non-negative integrable functions such that

$$h_1 - h_2 = h = h^+ - h^-,$$

then

$$\begin{aligned}&h_1 + h^- = h_2 + h^+, \\ &\int h_1 + \int h^- = \int h_2 + \int h^+, \\ &\int h_1 - \int h_2 = \int h^+ - \int h^- = \int h.\end{aligned}$$

Since $(f^+ + g^+)$ and $(f^- + g^-)$ are non-negative integrable functions whose difference is $f+g$, we have

$$\begin{aligned}\int f+g &= \int (f^+ + g^+) - \int (f^- + g^-) \\ &= \int f^+ + \int g^+ - \int f^- - \int g^- = \int f + \int g.\end{aligned}$$

(iii) If $f \le g$, then $g - f \ge 0$ so by (i) and (ii),

$$\int (g-f) = \int g - \int f,$$

and of course $\int(g - f) \geq 0$. ▥

Problem 9. Prove that $\int kf = k\int f$ if $k < 0$. ▥

Problem 10. If f is integrable over T, then $\int_S f = \int_T f\chi_S$ for every measurable set $S \subset T$. ▥

Problem 11. If A and B are disjoint measurable sets, and f is integrable over A and B, then $\int\limits_{A\cup B} f = \int\limits_A f + \int\limits_B f$. ▥

Proposition 5. *(Lebesgue Dominated Convergence Theorem). If $\{f_n\}$ is a sequence of measurable functions and $f_n \longrightarrow f$ a.e., and $|f_n| \leq g$ for some integrable g and all n, then*

$$\int f_n \longrightarrow \int f.$$

Proof. The convergence $f_n \longrightarrow f$ goes on entirely in the finite area bounded by g and $-g$, so the convergence of the integrals follows from Proposition 3. Specifically, $f_n^+ \longrightarrow f^+$ and $0 \leq f_n^+ \leq g$ for all n, so $\int f_n^+ \longrightarrow \int f^+$. Similarly, $f_n^- \longrightarrow f^-$ and $0 \leq f_n^- \leq g$ for all n, so $\int f_n^- \longrightarrow \int f^-$. ▥

We finish this chapter with the result that every integral can be approximated by Riemann sums.

Proposition 6. *Let f be integrable (over any measurable set S). For every $\varepsilon > 0$ there is a finite measure set T and a partition $\{E_i\}$ of T and a choice $c_i \in E_i$ so that*

$$\left|\int f - \sum f(c_i)\mu(E_i)\right| < \varepsilon.$$

Proof. It is sufficient to show this for non-negative functions f (Problem 12 below). Let $f \geq 0$ and let g be a primary function with $0 \leq g \leq f$, $g = 0$ off a finite measure set T, and

$$\int f - \varepsilon < \int_T g \leq \int f.$$

Choose a partition $P = \{E_i\}$ of T such that $\mu(E_i) > 0$ for each i and $U(g,P) - L(g,P) < \varepsilon$. Then

$$\int f - 2\varepsilon < L(g,P) \leq \int_T g \leq \int f.$$

Let

$$m_i = \inf\{g(x) : x \in E_i\},$$
$$\overline{m}_i = \inf\{f(x) : x \in E_i\}.$$

Since f cannot be identically $+\infty$ on a set of positive measure, each $\overline{m}_i < \infty$, and $0 \le m_i \le \overline{m}_i$ for all i. Hence

$$\int f - 2\varepsilon < L(g, P) = \sum m_i \mu(E_i) \le \sum \overline{m}_i \mu(E_i).$$

If $\varphi = \sum \overline{m}_i \chi_{E_i}$, then φ is a primary function under f, so

$$\int \varphi = \sum \overline{m}_i \mu(E_i) \le \int f.$$

Therefore

$$\int f - 2\varepsilon < \sum \overline{m}_i \mu(E_i) \le \int f.$$

Now choose $c_i \in E_i$ so

$$\overline{m}_i \le f(c_i) < \overline{m}_i + \varepsilon/\mu(T).$$

Then

$$\begin{aligned}\int f - 2\varepsilon &< \sum f(c_i)\mu(E_i) \\ &\le \sum \Big(\overline{m}_i + \varepsilon/\mu(T)\Big)\mu(E_i) \\ &= \sum \overline{m}_i \mu(E_i) + \varepsilon \\ &< \int f + \varepsilon.\end{aligned}$$

Hence

$$\left|\int f - \sum f(c_i)\mu(E_i)\right| < 2\varepsilon. \quad ▮$$

Corollary. *For every integrable function f there is a simple function $\sum y_i \chi_{E_i} = \varphi$ such that $|\varphi| \leq |f|$ and $\left|\int f - \int \varphi\right| < \varepsilon$.*

Problem 12. Prove Proposition 6 for functions f which are not necessarily non-negative.

Problem 13. (A generalization of the Dominated Convergence Theorem.) Another way to ensure that functions $\{f_n\}$ cannot reach out to enclose a mischievous amount of area beyond the limit function f is to require that each f_n be dominated by some g_n where the integrals $\int g_n$ converge. Prove: Let $\{f_n\}$, $\{g_n\}$ be sequences of non-negative integrable functions such that $f_n \longrightarrow f$ a.e., and $g_n \longrightarrow g$ a.e. If $f_n \leq g_n$ for all n, and $\int g_n \longrightarrow \int g < \infty$, then $\int f_n \longrightarrow \int f$. Hint: The Fatou Lemma is the basic fact about convergence of integrals.

Problem 14. Let $f_n(x) = n^{3/2}x/(1 + n^2x^2)$ for $x \in [0,1]$.

(a) Show that $f_n(x) \longrightarrow 0$ for all $x \in [0,1]$.

(b) Show that $\{f_n\}$ is not uniformly bounded on $[0,1]$.

(c) Show that $g(x) = 1/\sqrt{x}$ is integrable on $[0,1]$ (*i.e.*, Lebesgue integrable - not improperly Riemann integrable), and that $f_n \leq g$ for all n.

(d) Conclude (from what theorem) that $\int f_n \longrightarrow 0$.

Problem 15. We saw earlier that allowing countable partitions gains nothing for the case where the function is bounded and defined on a finite measure set. However, allowing countable partitions does allow one to define the integral for all positive functions at one fell swoop. That is, it is not necessary to consider first the primary functions, and then unbounded functions on infinite measure sets.

Let f be a positive, possibly unbounded, measurable function on $\mathbb{R}$. A countable partition of $\mathbb{R}$ is a countable family $P = \{E_i\}$ of disjoint measurable sets whose union is $\mathbb{R}$. Let M_i, m_i, $U(f,P)$, $L(f,P)$ be defined as usual. Say that f is integrable if and only if $\sup_P L(f,P) = \inf_P U(f,P) < \infty$, where now P ranges over countable partitions. Show that if f, which is measurable, has one finite upper sum, then f is integrable.

Problem 16. (a) Let f be a positive measurable function on

$\mathbb{R}$. Let $S_n = \{x : f(x) \geq 2^n\}$ for $n = 0, \pm 1, \pm 2, \ldots$. Let $E_n = S_n - S_{n+1}$ so $P = \{E_n : n = 0, \pm 1, \pm 2, \ldots\}$ is a (countable) partition of $\mathbb{R}$. Show that f is integrable if and only if $\sum_{-\infty}^{\infty} 2^n \mu(S_n) < \infty$. Hint: If the series converges, then

$$U(f,P) \leq \sum_{-\infty}^{\infty} 2^{n+1} \mu(E_n) = \sum_{-\infty}^{\infty} 2^n \mu(S_n).$$

Suppose the series diverges. If the sequence $\{2^{N+1}\mu(S_{N+1}) : N = 0, 1, 2, \ldots,\}$ is unbounded, f is clearly not integrable so suppose $2^{N+1}\mu(S_{N+1}) \leq B$ for all $N \geq 0$. Let M be a large number and pick N so that $\sum_{-N+1}^{N} 2^n \mu(S_n) > M$. Show that

$$L(f,P) \geq \sum_{-N}^{N} 2^n \mu(E_n) \geq \frac{1}{2}M - B.$$

(Notice that this is just the integral test of the calculus, only now the series is used to test "convergence" of the integral.)

(b) Now let $T_n = \{x : f(x) \geq r^n\}$, where $r > 1$ and r is close to 1. Let $E_n = \{x : r^n \leq f(x) < r^{n+1}\}$ for $n = 0, \pm 1, \pm 2, \ldots$, and let $P = \{E_n\}$. Assuming that $\sum_{-\infty}^{\infty} r^n \mu(T_n) < \infty$, show that

$$U(f,P) \leq (r-1) \sum_{-\infty}^{\infty} r^n \mu(T_n),$$

$$L(f,P) \geq (1 - \frac{1}{r}) \sum_{-\infty}^{\infty} r^n \mu(T_n).$$

Problem 17. (i) Let $\mu(E_n) < 1/2^n$ and let $g_n = \chi_{E_n}$. Show that $g_n \longrightarrow 0$ a.e. Hint: Let $F_N = \bigcup_{i=N}^{\infty} E_i$, and let $F = \bigcap_N F_N$. Show $\mu(F) = 0$ and $g_n(x) \longrightarrow 0$ if $x \notin F$.

(ii) Show that $\mu(E_n) \longrightarrow 0$ is not sufficient to ensure that $g_n \longrightarrow 0$ a.e. Hint: Let $E_1 = [0, \frac{1}{2}]$, $E_2 = [\frac{1}{2}, 1]$, $E_3 = [0, \frac{1}{4}]$, $E_4 = [\frac{1}{4}, \frac{1}{2}]$, $E_5 = [\frac{1}{2}, \frac{3}{4}]$, $E_6 = [\frac{3}{4}, 1]$, $E_7 = [0, \frac{1}{8}]$, etc.

Problem 18. If f is integrable on S then for every $\varepsilon > 0$ there is $\delta > 0$ so that $\int_E |f| < \varepsilon$ whenever $E \subset S$ and $\mu(E) < \delta$. Hint: Assume without loss that $f \geq 0$ and suppose the statement false. Then there is $\varepsilon > 0$ and a sequence $\{E_n\}$ of measurable subsets of S with $\mu(E_n) < 1/2^n$ and $\int_{E_n} f \geq \varepsilon$. Use Problem 17 and the Lebesgue Dominated Convergence Theorem with $g_n = \chi_{E_n} \cdot f$ to reach a contradiction. ◀

Problem 19. (Differentiation under the integral sign.) Let $f(x,t)$ be defined on $[a,b] \times (c,d)$, with $t_0 \in (c,d)$. Assume that for each fixed $t \in (c,d)$, $f(x,t)$ is a measurable function of x on $[a,b]$, and for each fixed $x \in [a,b]$, $f(x,t)$ is a differentiable function of t on (c,d); *i.e.*, $f_t(x,t)$ exists for $(x,t) \in [a,b] \times (c,d)$. Assume that $f(x,t_0)$ is integrable on $[a,b]$, and that there is an integrable function g on $[a,b]$ such that $|f_t(x,t)| \leq g(x)$ for all $(x,t) \in [a,b] \times (c,d)$. Show that:

(i) $f_t(x,t)$ is a measurable function of x for each $t \in (c,d)$, and hence $f_t(x,t)$ is integrable on $[a,b]$.

(ii) For each $t \in (c,d)$, $f(x,t)$ is integrable for $x \in [a,b]$.

(iii) If $F(t) = \int_a^b f(x,t)d\mu(x)$ for $t \in (c,d)$, then $F'(t_0) = \int_a^b f_t(x,t_0)d\mu(x)$.

Hints: (i) Let $t_n \longrightarrow t \in (c,d)$, so for $x \in [a,b]$

$$\frac{f(x,t_n) - f(x,t)}{t_n - t} \longrightarrow f_t(x,t)\,.$$

(ii) For each $t \in (c,d)$ there is $s \in (c,d)$ so $f(x,t) = f(x,t_0) + f_t(x,s)(t-t_0)$, and hence $|f(x,t)| \leq |f(x,t_0)| + |f_t(x,s)||t-t_0|$.

(iii) Let $t_n \longrightarrow t_0$, so

$$F'(t_0) = \lim_n \int_a^b \frac{f(x,t_n) - f(x,t_0)}{t_n - t_0} d\mu(x)\,.$$

The difference quotients are dominated by $g(x)$ and converge pointwise to $f_t(x,t_0)$. ◀

8 DIFFERENTIATION AND INTEGRATION

In this chapter we make some connections between differentiation and integration. In particular, we see how the Fundamental Theorem of Calculus fares with respect to Lebesgue integration. The results of this chapter are necessarily germane only to real valued functions defined on the real line. This is in contrast to our earlier work, where the results are stated for functions and sets on the line, but the techniques and proofs have quite general application. The theorems of this chapter are standard useful results, but the proofs tend to be difficult and of limited applicability. The reader may be well-advised to study the statements of the theorems and problems, and leave the proofs for a long rainy day.

The two forms of the Fundamental Theorem are

$$\frac{d}{dx}\int_a^x f = f(x), \tag{1}$$

$$\int_a^b f' = f(b) - f(a). \tag{2}$$

In calculus one shows that (1) holds for a Riemann integrable function f which is continuous at x. This is also true for a Lebesgue integrable function, with essentially the same proof (Problem 1). It is considerably more difficult (Proposition 5) to show that if f is bounded and measurable, and so Lebesgue integrable, then (1) holds a.e.

Problem 1. Show that if f is Lebesgue integrable on an interval containing a and x_0, and f is continuous at x_0, and $a < x_0$, and

$$F(x) = \int_a^x f,$$

then $F'(x_0) = f(x_0)$. ◀

We will now show that (2) does not hold in general by constructing a continuous increasing function f on $[0,1]$ with $f(0) = 0$, $f(1) = 1$, and $f'(x) = 0$ a.e. Thus

$$\int_0^1 f' = 0 \neq f(1) - f(0) = 1\,.$$

Define f on $[0,1]$ as follows: let $f = \frac{1}{2}$ on $\left[\frac{1}{3}, \frac{2}{3}\right]$; then let $f = \frac{1}{4}$ on $\left[\frac{1}{9}, \frac{2}{9}\right]$ and $f = \frac{3}{4}$ on $\left[\frac{7}{9}, \frac{8}{9}\right]$. In the next step, take the middle third out of each of the four remaining intervals, and define f to be $\frac{1}{8}, \frac{3}{8}, \frac{5}{8}$, and $\frac{7}{8}$ on these middle thirds. The picture this far is Fig. 1.

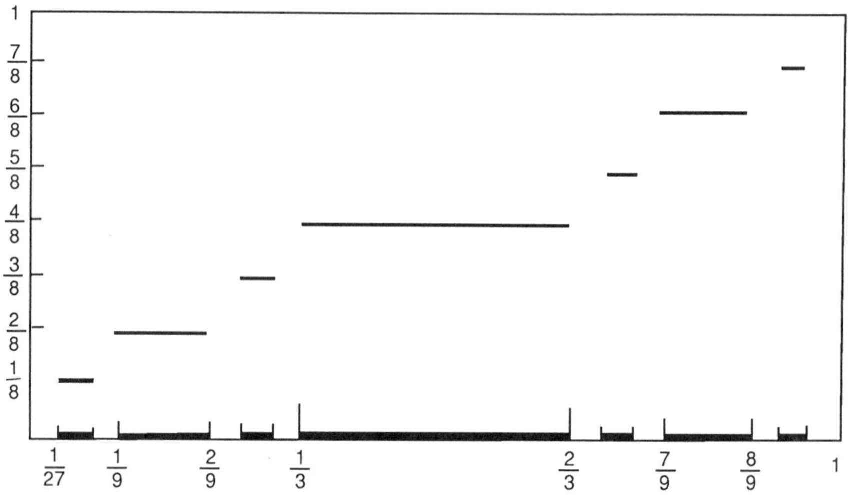

Fig.1

Continuing in this way we get f defined on a union of disjoint closed intervals with total length

$$\frac{1}{3} + \frac{2}{9} + \frac{4}{27} + \cdots = 1\,.$$

Clearly $f'(x) = 0$ on the union of the interiors of these intervals, so $f'(x) = 0$ a.e. The inductive scheme outlined above defines f on a dense subset of $[0,1]$. To show that this definition can be extended to the rest of $[0,1]$ with the result being continuous, it suffices (Problem 2) to show that f is uniformly continuous on the union of the closed intervals. Notice that

$$\text{if } |x_1 - x_2| < \frac{1}{9}, \text{ then } |f(x_1) - f(x_2)| < \frac{1}{4},$$
$$\text{if } |x_1 - x_2| < \frac{1}{27}, \text{ then } |f(x_1) - f(x_2)| < \frac{1}{8},$$

and so on, so f is uniformly continuous on its domain. The function f is known as the **Cantor singular function.**

Problem 2. Let f be defined and uniformly continuous on a dense subset E of $[0,1]$. Specifically, given $\varepsilon > 0$ there is $\delta > 0$ so that if $x_1, x_2 \in E$ and $|x_1 - x_2| < \delta$, then $|f(x_1) - f(x_2)| < \varepsilon$. Let $x \in [0,1]$ and let $\{x_n\}$ be a sequence in E such that $x_n \longrightarrow x$. (x may or may not be in E.) Show that $\{f(x_n)\}$ converges, and the limit is independent of the sequence $\{x_n\}$ converging to x. If $x \in E$, show that $\lim f_n(x) = f(x)$. If $x \notin E$, define $f(x) = \lim f_n(x)$, where $x_n \in E$ and $x_n \longrightarrow x$. Show that f is continuous on $[0,1]$. ◀

A **Vitali covering** of a set E is a family V of proper intervals (open or closed or half-open, but not points) such that every point of E lies in intervals of V of arbitrarily small length. That is, given $x \in E$ and $\eta > 0$ there is $I \in V$ such that $x \in I$ and $\ell(I) < \eta$. The proof that an increasing function has a derivative a.e. depends on the following proposition.

Proposition 1. *(Vitali's Theorem). If $\mu(E) < \infty$ and V is a Vitali covering of E, then for each $\varepsilon > 0$ there is a finite disjoint set $\{I_1, \dots, I_N\}$ of intervals of V such that $\mu(E - (I_1 \cup \cdots \cup I_N)) < \varepsilon$, and hence $\mu\left(E \cap \bigcup I_j\right) > \mu(E) - \varepsilon$.*

Proof. Let U be an open set containing E with $\mu(U) < \infty$. Discard from V all intervals not contained in U, and all intervals which do not intersect E. What remains is of course still a Vitali covering of E. We can assume without loss that all intervals of V are closed, because closing them would affect neither the hypothesis nor the conclusion. Suppose we let I_1 be any interval from V of maximum length, and I_2 any interval of maximum length which does not intersect I_1. Of course there may not be such maximum length intervals, but for the moment we assume there are to see how the argument goes. Later we will correct the argument by replacing these by intervals which are nearly of maximal length. Define inductively a disjoint sequence $I_1, I_2, \ldots$ of intervals of V such that each I_{n+1} is of largest possible length and disjoint from $I_1, \ldots, I_n$. If at any stage E is covered by $I_1 \cup \cdots \cup I_n$ we are done. Otherwise, since $I_1 \cup \cdots \cup I_n$ is closed, each $x \in E-(I_1 \cup \cdots \cup I_n)$ is in some intervals of V which are so small they miss $I_1 \cup \cdots \cup I_n$. Thus we have a sequence $\{I_n\}$ of disjoint intervals, all contained in U, with $\sum \ell(I_n) < \infty$ and $\ell(I_n) \longrightarrow 0$. Let $\varepsilon > 0$ and pick N so large that $\sum_{N+1}^{\infty} \ell(I_n) < \varepsilon$. Consider any point x of $E-(I_1 \cup \cdots \cup I_N)$ and any interval I of V which contains x and is disjoint from $I_1 \cup \cdots \cup I_N$. The interval I must intersect some I_m with $m > N$, for otherwise since I_m is of maximal length among unchosen intervals, $\ell(I_m) \geq \ell(I)$ for all m, and this is impossible since $\ell(I_m) \longrightarrow 0$. Let I_m be the first interval that I intersects, so $m > N$. Since $\ell(I_m) \geq \ell(I)$, x lies in the interval J_m which has the same center as I_m but is three times as long(Fig. 2).

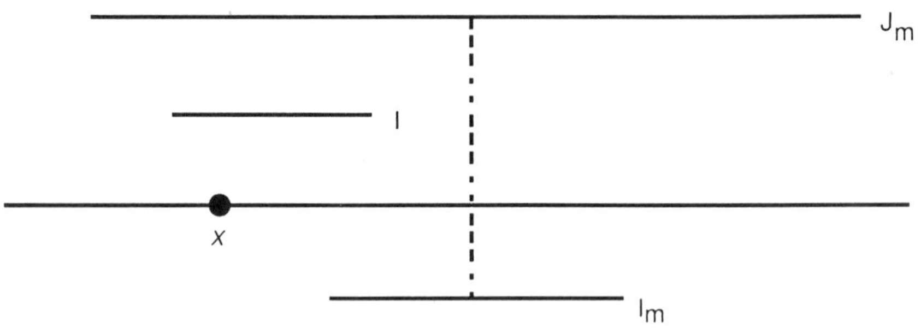

$x \in J_m$; $\ell(J_m) = 3\ell(I_m) \geq 3\ell(I)$.

Fig. 2

Hence every point x of $E \sim (I_1 \cup \cdots \cup E_N)$ is in $\bigcup_{m=N+1}^{\infty} J_m$, which has measure less than 3ε.

Now instead of assuming that each I_{n+1} has maximum length, let $L_n = \sup\{\ell(I) : I \in V \text{ and } I \cap (I_1 \cup \cdots \cup I_n) = \emptyset\}$. Then pick I_{n+1} so that $\ell(I_{n+1}) \geq \frac{1}{2}L_n$. The arguments above all proceed as before except we must make J_m five times as long as I_m, so that $E - (I_1 \cup \cdots \cup I_N)$ has measure less than 5ε. ▥

Problem 3. Let q be any function and assume that

$$\{x : \liminf_{t \to x+} q(t) < \limsup_{t \to x+} q(t)\}$$

has positive (outer) measure. Show there are rationals r and s such that

$$\{x : \liminf_{t \to x+} q(t) < r < s < \limsup_{t \to x+} q(t)\}$$

has positive (outer measure). Hint: The measurability of the sets is not relevant here. ◀

Proposition 2. *If f is increasing on $[a,b]$ then $f'(x)$ exists a.e.*

Proof. The four **Dini derivates** of f at x are:

$$D^+f(x) = \limsup_{h \to 0^+} \frac{f(x+h) - f(x)}{h},$$

$$D_+f(x) = \liminf_{h \to 0^+} \frac{f(x+h) - f(x)}{h},$$

$$D^-f(x) = \limsup_{h \to 0^-} \frac{f(x+h) - f(x)}{h},$$

$$D_-f(x) = \liminf_{h \to 0^-} \frac{f(x+h) - f(x)}{h}.$$

The derivative $f'(x)$ exists if and only if the four derivates are equal and finite at x. The proof consists in showing that the set where any two derivates differ has measure zero. We illustrate by showing that the set $\{x : D^+f(x) > D_+f(x)\}$ has measure zero. If this set has positive measure, then there are (Problem 3) rational numbers r and s such that

$$A = \{x : D_+ f(x) < r < s < D^+ f(x)\}$$

has positive measure. Assume $\mu(A) = p > 0$, and let U be an open superset of A so that $\mu(U) < p + \varepsilon$. For each $x \in A$ there are, by definition of $D_+ f(x)$, arbitrarily small numbers h such that $[x, x+h] \subset U$ and

$$f(x+h) - f(x) < rh\,.$$

These intervals $[x, x+h]$ form a Vitali covering of A, so there is a finite disjoint collection $[x_i, x_i + h_i]$, $i = 1, 2, \ldots, N$, with

$$\mu\left(A \cap \bigcup [x_i, x_i + h_i]\right) > p - \varepsilon\,.$$

Since $\bigcup [x_i, x_i + h_i] \subset U$,

$$\sum_{i=1}^{N} h_i < \mu(U) < p + \varepsilon\,.$$

Let $B = A \cap \bigcup_{i=1}^{N} (x_i, x_i + h_i)$, so that $\mu(B) > p - \varepsilon$ and each point of B is the left endpoint of an interval $[y, y+k]$, contained in some $(x_i, x_i + h_i)$, such that

$$f(y+k) - f(y) > sk\,.$$

This last inequality follows from the fact that $D^+ f(x) > s$ for each $x \in B$. Invoke Vitali's Theorem again to get a finite disjoint family $[y_j, y_j + k_j]$, $j = 1, \ldots, M$, such that

$$\mu\left(B \cap \bigcup_{j=1}^{M} [y_j, y_j + k_j]\right) > \mu(B) - \varepsilon > p - 2\varepsilon\,.$$

It follows that $\sum_{j=1}^{M} k_j > p - 2\varepsilon$. Now we have

$$\sum_{i=1}^{N} (f(x_i + h_i) - f(x_i)) < r \sum_{i=1}^{N} h_i < r(p + \varepsilon)\,,$$

and

$$\sum_{j=1}^{M}(f(y_j + k_j) - f(y_j)) > s\sum_{j=1}^{M} k_j > s(p - 2\varepsilon)\,.$$

Since each interval $[y_j, y_j + k_j]$ is contained in some $[x_i, x_i + h_i]$, and f is increasing,

$$\sum_{j=1}^{M}(f(y_j + k_j) - f(y_j)) \le \sum_{i=1}^{N}(f(x_i + h_i) - f(x_i))\,.$$

Thus, for every $\varepsilon > 0$

$$s(p - 2\varepsilon) < r(p + \varepsilon),$$

so

$$\begin{aligned} sp &\le rp, \\ s &\le r, \end{aligned}$$

and we have a contradiction. ▮

Proposition 3. *If f is increasing on $[a, b]$, then f' is measurable and $\int_a^b f' \le f(b) - f(a)$.*

Proof. Let

$$f_n(x) = n\left[f\left(x + \frac{1}{n}\right) - f(x)\right]$$

where we interpret $f(x + \frac{1}{n}) = f(b)$ if $x + \frac{1}{n} \ge b$. Then f_n is measurable for each n, and $f_n(x) \longrightarrow f'(x)$ a.e., so f' is measurable. Each f_n is non-negative, so Fatou's Lemma gives

$$\begin{aligned} \int_a^b f' &= \int_a^b \lim f_n \\ &\le \liminf \int_a^b f_n \\ &= \liminf \left(n\int_b^{b+\frac{1}{n}} f - n\int_a^{a+\frac{1}{n}} f \right) \\ &= f(b) - \limsup n\int_a^{a+\frac{1}{n}} f \\ &\le f(b) - f(a)\,. \end{aligned}$$

The last inequality follows from the fact that f is increasing, so for all n,

$$n \int_a^{a+\frac{1}{n}} f \geq n f(a) \frac{1}{n} = f(a) . \quad ▥$$

Proposition 4. *If f is integrable on $[a, b]$ and $\int_0^x f = 0$ for all $x \in [a, b]$, then $f = 0$ a.e.*

Proof. Suppose $\int_a^x f = 0$ for all x, so $\int_c^d f = 0$ for all c, $d \in (a, b)$. Suppose, to be specific, that f is positive on a set E of positive measure. Then there is a closed subset $F \subset E$ so that $\mu(F) > 0$ and $\int_F f > 0$. Let U be the open set $(a, b) - F$, and write U in terms of its components: $U = \bigcup(a_i, b_i)$. Since $U \cup F = (a, b)$,

$$\int_U f = -\int_F f < 0 .$$

Therefore

$$\sum_i \int_{a_i}^{b_i} f < 0 ,$$

and so $\int_{a_i}^{b_i} f \neq 0$ for some interval (a_i, b_i), which contradicts the assumption. ▥

The following result is true for functions which are integrable on $[a, b]$, and so not necessarily bounded. The proof of the more general theorem requires additional machinery, and so we stick with the version below, which contains the essential ideas.

Proposition 5. *If f is bounded and measurable on $[a, b]$ and*

$$F(x) = \int_a^x f ,$$

then F is continuous and $F(a) = 0$ and $F'(x) = f(x)$ a.e. (The points where $F'(x) = f(x)$ are called **Lebesgue points** *for f.)*

Proof. Let $f^+ = f \vee 0$ and $f^- = (-f) \vee 0$, so f^+ and f^- are non-negative functions and $f = f^+ - f^-$. Hence

$$F(x) = \int_a^x f^+ - \int_a^x f^- = F_1(x) - F_2(x) ,$$

where F_1, F_2 are increasing functions. It follows from Proposition 2 that $F'(x)$ exists for almost all x. Moreover, if M is a bound for f,

$$|F(x_2) - F(x_1)| = \left| \int_{x_1}^{x_2} f \right| \le M|x_2 - x_1|,$$

so F is continuous. Let

$$\begin{aligned} f_n(x) &= n\left[F\left(x + \frac{1}{n}\right) - F(x)\right] \\ &= n\int_x^{x+\frac{1}{n}} f, \end{aligned}$$

so that $|f_n(x)| \le M$ for all x, and $f_n(x) \longrightarrow F'(x)$ a.e. By the Bounded Convergence Theorem and the continuity of F,

$$\begin{aligned} \int_a^x F' &= \lim_n \int_a^x f_n \\ &= \lim_n \left[\int_a^x nF\left(t + \frac{1}{n}\right) dt - \int_a^x nF(t)dt\right] \\ &= \lim_n \left[\int_x^{x+\frac{1}{n}} nF(t)dt - \int_a^{a+\frac{1}{n}} nF(t)dt\right] \\ &= F(x) - F(a) \\ &= F(x) = \int_a^x f. \end{aligned}$$

Now we have

$$\int_a^x (F' - f) = 0$$

for all x, and consequently $F' = f$ a.e. by Proposition 4. ▮

The Lebesgue points play critical roles in many theorems on integral representations. We give one example from potential theory. The problem is the so-called Dirichlet problem; namely, to find a harmonic function $u(r, \theta)$ on the open unit disc in the plane with a specified boundary value $f(\varphi)$ on the unit circle. Here we will let f be a bounded measurable function on $[0, 2\pi]$, which we regard as the unit circle. The Poisson kernel is the function

$$P(r, \theta; \varphi) = \frac{1 - r^2}{1 + r^2 - 2r\cos(\theta - \varphi)},$$

defined for (r, θ) in the open unit disc (polar coordinates), and φ on the unit circle. For each fixed $\varphi_0, P(r, \theta; \varphi_0)$ is a harmonic function on the open disc $(0 \le r < 1, 0 \le \theta \le 2\pi)$. For each fixed (r, θ), $P(r, \theta; \varphi)$ is a continuous function of $\varphi \in [0, 2\pi]$. Hence $f(\varphi)P(r, \theta; \varphi)$ is an integrable function of φ for each (r, θ) in the disc. We let

$$u(r, \theta) = \int_0^{2\pi} f(\varphi)P(r, \theta; \varphi)\frac{1}{2\pi}d\mu(\varphi).$$

The $\frac{1}{2\pi}$ is so the total measure of the circle is 1. We can regard the integral as a limit of Riemann sums, so

$$u(r, \theta) = \lim \sum_{i=1}^{N} P(r, \theta; \varphi_i)f(\varphi_i)\frac{1}{2\pi}\mu(E_i).$$

The sums have the form

$$\sum_{i=1}^{N} a_i p_i(r, \theta) \tag{3}$$

where $p_i(r, \theta) = P(r, \theta; \varphi_i)$ and $a_i = \frac{1}{2\pi}f(\varphi_i)\mu(E_i)$. The sums (3) are linear combinations of harmonic functions, and hence are harmonic. The limit of these sums, $u(r, \theta)$, is also harmonic, although not obviously so. The function $u(r, \theta)$ satisfies the Dirichlet problem in the following sense: if φ_0 is a Lebesgue point for f (and that includes all points where f is continuous) then

$$\lim_{r \to 1-} u(r, \varphi_0) = f(\varphi_0). \tag{4}$$

In particular, (4) holds for almost every φ_0.

Problem 4. Let $F(x) = \int_0^x f$. Show that $F'(0)$ can exist even though f is not continuous at 0. Hint: Let f be defined on $[0, 1]$ as follows: $f = 1$ on $[\frac{1}{2}, \frac{3}{4}]$ and $f = -1$ on $(\frac{3}{4}, 1]$, so $0 \le \int_{\frac{1}{2}}^x f \le \frac{1}{4}$ for $\frac{1}{2} \le x \le 1$. Divide $[\frac{1}{4}, \frac{1}{2})$ up into four equal intervals, and let f be alternately $+1$ and -1 on these, so $0 \le \int_{\frac{1}{4}}^x f \le \frac{1}{16}$ for $\frac{1}{4} \le x \le \frac{1}{2}$, etc. Show that f is integrable - even properly Riemann integrable since the set of discontinuities has measure zero - and that $(F(x) - F(0))/x \longrightarrow 0$ as $x \longrightarrow 0+$. ◾

9 PLANE MEASURE

In this chapter we develop Lebesgue measure for sets in the plane, $\mathbb{R}^2$. Our purpose is three-fold. First, by developing another specific example of a measure we show how the same techniques used on the line can be used to define countably additive measures in quite general situations. Our second purpose is to have two dimensional measure - area - defined so we can show that $\int f d\mu$ really is the area under the graph of f. Thirdly, the development of plane measure provides a template and example for defining general product measures.

We take the rectangle as our basic plane figure, with the rectangle playing the role played by the interval on the line. We will here use "rectangle" to mean rectangle with sides parallel to the axes, and having positive length and width. Thus a rectangle is a set $I \times J$ with I and J intervals, which may be open or closed or half-open. Single points and line segments are not rectangles. If $R = I \times J$, then the area of R is $\alpha(R) = \ell(I)\ell(J)$.

For any set $E \subset \mathbb{R}^2$,

$$\lambda(E) = \inf \left\{ \sum \alpha(R_i) : E \subset \bigcup R_i \right\}$$

where $\{R_i\}$ is a finite or countable family of rectangles. $\lambda(E)$ is the **Lebesgue outer measure** of E which we will call simply the **measure** of E.

The same arguments used when we defined $m(E)$ on the line will show that it is immaterial whether we use coverings of E by open rectangles or closed rectangles or partially closed rectangles (Problem 2). We can also invoke the plane version of the Heine-Borel Theorem and notice that if E is a compact set, then $\lambda(E)$ is the inf of sums $\sum \alpha(R_i)$ for **finite** coverings of E by rectangles of any sort.

Problem 1. (i) Show that line segments have plane measure zero; for example, the segment from $(0,0)$ to $(1,2)$ has measure zero.

(ii) Show that lines have plane measure zero.

It will occasionally be convenient to use coverings by squares of the form $d \times e$, where $d = \left[\frac{k}{2^n}, \frac{k+1}{2^n}\right]$, $e = \left[\frac{\ell}{2^n}, \frac{\ell+1}{2^n}\right]$. We will call an interval of the form $\left[\frac{k}{2^n}, \frac{k+1}{2^n}\right]$, with or without endpoints, a **dyadic interval** of length $\frac{1}{2^n}$. A square of the form $d \times e$ will be called a **dyadic square** of side $\frac{1}{2^n}$.

Any rectangle R can obviously be covered by a finite number of dyadic squares, $S_1, S_2, \ldots, S_n$, with

$$\sum_{j=1}^{n} \alpha(S_j) < \alpha(R) + \varepsilon .$$

Any countable covering $\{R_n\}$ of any set can be covered by countably many dyadic squares $\{S_{nj}\}$, where $S_{n1}, S_{n2}, \ldots$ cover R_n, and

$$\sum_{j} \alpha(S_{nj}) < \alpha(R_n) + \varepsilon/2^n ,$$

so

$$\sum_{j,n} \alpha(S_{nj}) < \sum \alpha(R_n) + \varepsilon .$$

Thus we could replace coverings of E by rectangles in our definition of λ by countable coverings of E by dyadic squares. Notice that a finite covering of E by non-overlapping dyadic squares can be replaced by a finite covering of non-overlapping dyadic squares all of the same size, with the same total area. (Rectangles are called **non-overlapping** provided their interiors are disjoint.)

Problem 2. Show that the infimum of sums $\sum \alpha(R_i)$ for families $\{R_i\}$ of closed rectangles covering E is the same as the infimum of sums $\sum \alpha(Q_i)$ for coverings $\{Q_i\}$ of E by open rectangles, and the same as the infimum of sums $\sum \alpha(S_i)$ for coverings $\{S_i\}$ of E by dyadic squares.

Problem 3. λ is translation invariant.

The following elementary properties are immediate from the definition.

Proposition 1. *(i)* $\lambda(\varnothing) = 0$;
(ii) $\lambda(C) = 0$ *for every countable set* C;
(iii) $\lambda(E) \geq 0$ *for all* E;
(iv) if $E \subset F$, *then* $\lambda(E) \leq \lambda(F)$;
(v) λ *is countably subadditive; i.e., if* $\{E_i\}$ *is any countable or finite family, then* $\lambda\left(\bigcup E_i\right) \leq \sum \lambda(E_i)$.

Problem 4. Prove parts (ii) and (v) of Proposition 1.

Since $\lambda(E)$ is to represent the area of E we need to know that λ gives the right answer for rectangles.

Proposition 2. *If* Q *is a rectangle,* $\lambda(Q) = \alpha(Q)$.

Proof. First assume that Q is a closed rectangle. Clearly $\lambda(Q) \leq \alpha(Q)$ since $\{Q\}$ is a 1-rectangle covering of itself. Fix $\varepsilon > 0$ and let $\{S_k\}$ be a finite covering of Q by dyadic squares of the same size so that $\sum \alpha(S_k) < \lambda(Q) + \varepsilon$. We may assume all S_k intersect Q. Since Q is a rectangle, say $Q = I \times J$, the squares S_k will consist of all squares $d_i \times e_j (i = 1, \dots, n;\ j = 1, \dots, m)$ where the d_i form a non-overlapping covering of I and the e_j form a non-overlapping covering of J. Hence

$$\begin{aligned}\bigcup_k S_k &= \bigcup_{i,j} d_i \times e_j \\ &= (d_1 \cup \cdots \cup d_n) \times (e_1 \cup \cdots \cup e_m),\end{aligned}$$

and

$$\begin{aligned}\lambda(Q) + \varepsilon &> \sum_{i,j} \alpha(d_i \times e_j) \\ &= \sum_i \ell(d_i) \sum_j \ell(e_j) \\ &\geq \mu(I)\mu(J).\end{aligned}$$

Since this holds for all $\varepsilon > 0$, and we already have $\lambda(Q) \leq \mu(I)\mu(J)$, we conclude that $\lambda(Q) = \mu(I)\mu(J)$. The remaining cases, where Q is not necessarily closed, are left to the reader as an exercise. ▮

Problem 5. Complete the proof of Proposition 2 by showing that $\lambda(Q) = \alpha(Q)$ for an open rectangle Q, and consequently, by monotonicity, for rectangles containing some but not all of their boundary points. ◀

We saw in Chapter 4 that the Carathéodory criterion for measurability, while not in itself very intuitive, does get us speedily to the additivity properties that outer measure has when restricted to the measurable sets. We therefore adopt this condition forthwith as our *definition* of *measurability.*

A set $E \subset \mathbb{R}^2$ is **measurable** if and only if

$$\lambda(E \cap T) + \lambda(E' \cap T) = \lambda(T) \tag{1}$$

for every set T. Here $E' = \mathbb{R}^2 - E$.

Since λ is subadditive, we always have

$$\lambda(E \cap T) + \lambda(E' \cap T) \geq \lambda(T),$$

so E is measurable if and only if, for all T,

$$\lambda(E \cap T) + \lambda(E' \cap T) \leq \lambda(T). \tag{2}$$

Clearly we need only consider sets T of finite measure.

Problem 6. (i) Sets of measure zero are measurable.

(ii) If $E_2 = E_1 \cup E_0$ with E_1 measurable and $\lambda(E_0) = 0$, then E_2 is measurable. ◀

Problem 7. Translates of measurable sets are measurable. ◀

Proposition 3. *E is measurable if and only if*

$$\lambda(E \cap R) + \lambda(E' \cap R) = \lambda(R)$$

for every rectangle R.

Proof. (Cf. the proof of Proposition 4, Chapter 4). The condition is obviously necessary, so assume that E splits all rectangles additively, and let T be any set of finite measure. Let $\{R_j\}$ be a covering of T by rectangles with

$$\sum \alpha(R_j) < \lambda(T) + \varepsilon.$$

Then $E \cap T \subset \bigcup(E \cap R_j)$ and $E' \cap T \subset \bigcup(E' \cap R_j)$. Hence, by monotonicity and subadditivity,

$$\begin{aligned} \lambda(E \cap T) + \lambda(E' \cap T) &\le \sum \lambda(E \cap R_j) + \sum \lambda(E' \cap R_j) \\ &= \sum [\lambda(E \cap R_j) + \lambda(E' \cap R_j)] \\ &= \sum \lambda(R_j) < \lambda(T) + \varepsilon . \end{aligned}$$

Therefore $\lambda(E \cap T) + \lambda(E' \cap T) \le \lambda(T)$ and E is measurable. ▮

Proposition 4. *Rectangles are measurable.*

Proof. Let Q be the rectangle that we want to show is measurable. We show that Q splits any rectangle R additively. R is the union of $Q \cap R$ and at most eight other non-overlapping rectangles $S_1, \ldots, S_8$ whose areas total $\alpha(R)$. (See Fig. 1).

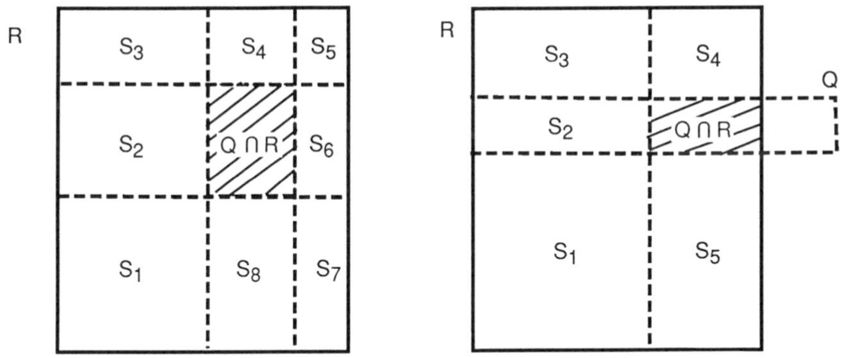

Fig. 1

Since $Q' \cap R \subset S_1 \cup \cdots \cup S_8$,

$$\lambda(Q' \cap R) \le \alpha(S_1) + \cdots + \alpha(S_8).$$

Therefore,

$$\begin{aligned}\lambda(Q \cap R) + \lambda(Q' \cap R) &\le \alpha(Q \cap R) + \alpha(S_1) + \cdots + \alpha(S_8)\\ &= \alpha(R) = \lambda(R).\end{aligned}$$

Subadditivity gives the other inequality, so Q splits any R additively, and hence Q is measurable. ||||||

Proposition 5. *If $\{E_i\}$ is a finite or countable family of disjoint measurable sets, then $\lambda(\bigcup E_i) = \sum \lambda(E_i)$.*

Proof. The proof of Proposition 5, Chapter 4 applies verbatim to this case. ||||||

Problem 8. If T is any set and $\{E_i\}$ is a finite or countable family of disjoint measurable sets, then

$$\lambda\left(T \cap \bigcup E_i\right) = \sum \lambda(T \cap E_i).$$

Proposition 6. *If $E_1, \ldots, E_n$ are measurable, then $E_1 \cup \cdots \cup E_n$ is measurable, $E_1 \cap \cdots \cap E_n$ is measurable, and $E_1 - E_2$ is measurable.*

Problem 9. Prove Proposition 6. Hint: Cf. Proposition 6, Chapter 4.

Proposition 7. *If $\{E_i\}$ is a countable family of measurable sets, then $\bigcup E_i$ is measurable.*

Proof. We can assume the sets are disjoint because the differences $E_2 - E_1$, $E_3 - (E_1 \cup E_2)$, etc. are measurable by the preceding proposition. If $F_n = E_1 \cup \cdots \cup E_n$ and $E = \bigcup_{i=1}^{\infty} E_i$, then for any T,

$$\begin{aligned}\lambda(T) &= \lambda(T \cap F_n) + \lambda(T \cap F_n')\\ &= \sum_{i=1}^{n} \lambda(T \cap E_i) + \lambda(T \cap F_n')\\ &\ge \sum_{i=1}^{n} \lambda(T \cap E_i) + \lambda(T \cap E').\end{aligned}$$

Since this holds for all n,

$$\begin{aligned}\lambda(T) &\geq \sum_{i=1}^{\infty} \lambda(T \cap E_i) + \lambda(T \cap E') \\ &\geq \lambda(T \cap E) + \lambda(T \cap E'). \end{aligned}$$

Problem 10. If A and B are measurable, $A \subset B$, and $\lambda(B) < \infty$, then $B - A$ is measurable and $\lambda(B - A) = \lambda(B) - \lambda(A)$.

Proposition 8. *Every open set in $\mathbb{R}^2$ is a countable union of open squares, so every open set and every closed set is measurable.*

Proof. Every point (x, y) of an open set U lies in arbitrarily small dyadic squares. Since some open disc around (x, y) is contained in U, a sufficiently small dyadic square containing (x, y) will lie in U. The union of all these squares is obviously U, and there are only a countable number of dyadic squares altogether. Squares are measurable, so open sets are measurable. If F is closed, then $F' = U$ is open, so measurable, and hence $F = U'$ is measurable.

Problem 11. For any measurable set E,

$$\begin{aligned}\lambda(E) &= \inf\{\lambda(U) : E \subset U, U \text{ open}\}. \\ &= \sup\{\lambda(F) : F \subset E, F \text{ compact}\}.\end{aligned}$$

Hint: Show this first for a set $E \subset [0, 1] \times [0, 1]$ and then write E as a countable union of sets $E \cap S_n$ where the S_n are non-overlapping unit squares.

Any measure ν on a topological space X is called **regular** provided that all open and closed sets are measurable, and for every measurable set E,

$$\begin{aligned}\nu(E) &= \inf\{\nu(U) : E \subset U, U \text{ open}\}, \\ \nu(E) &= \sup\{\nu(F) : F \subset E, F \text{ compact}\}.\end{aligned}$$

Problem 11 is the statement that λ is regular on $\mathbb{R} \times \mathbb{R}$. We have already seen that μ is regular on $\mathbb{R}$.

Problem 12. If for each $\varepsilon > 0$ there are measurable sets A and B such that $A \subset E \subset B$ and $\lambda(B) < \lambda(A) + \varepsilon$, then E is measurable. Show that the condition $\lambda(B) \leq \lambda(A) + \varepsilon$, which allows $\lambda(A) = \lambda(B) = \infty$, does not suffice. ◀

10 THE RELATIONSHIP BETWEEN μ AND λ

Since $\mathbb{R}^2 = \mathbb{R} \times \mathbb{R}$ there is naturally a relationship between plane measure and linear measure. In this chapter we will investigate the connection, and thereby preview the development of general product measures.

Proposition 1. *If A and B are subsets of $\mathbb{R}$ of finite outer measure (but not necessarily measurable sets), then $\lambda(A \times B) \le \mu(A)\,\mu(B)$.*

Proof. Let $\varepsilon > 0$ and let $\{I_j\}$, $\{J_k\}$ be coverings of A and B by intervals, with

$$\sum \ell(I_j) < \mu(A) + \varepsilon\,,$$
$$\sum \ell(J_k) < \mu(B) + \varepsilon\,.$$

Since $\{I_j \times J_k\}$ is a covering of $A \times B$ by rectangles, we have

$$\begin{aligned}\lambda(A \times B) &\le \sum_{j,k} \alpha(I_j \times J_k)\\ &= \sum_j \ell(I_j) \sum_k \ell(J_k)\\ &< (\mu(A) + \varepsilon)\,(\mu(B) + \varepsilon)\,.\end{aligned}$$

This holds for all $\varepsilon > 0$, and $\mu(A)$, $\mu(B)$ are finite, so $\lambda(A \times B) \le \mu(A)\,\mu(B)$. ▮

Problem 1. Let A, B be subsets of $\mathbb{R}$ with $\mu(A) = 0$. Show that $\lambda(A \times B) = 0$. Hint: If $\mu(B) < \infty$ this is clear. Otherwise, write $B = \bigcup B_n$ where $B_n = B \cap [n, n+1)$, $n = 0, \pm 1, \ldots$. ◂

Proposition 2. *If A and B are compact subsets of $\mathbb{R}$, then* $\lambda(A \times B) = \mu(A)\,\mu(B)$.

Proof. Since A and B are closed and bounded sets they are automatically measurable with respect to μ, and have finite measure. By Proposition 1 we need only show that $\lambda(A \times B) \geq \mu(A)\,\mu(B)$.

Since $A \times B$ is compact we can approximate $\lambda(A \times B)$ by the total areas of finite coverings by dyadic squares all of the same size. Let $\{S_{ij}\} = \{e_i \times d_j\}$ be such a covering, where all e_i and d_j are dyadic intervals of length $1/2^N$, and

$$\sum_{i,j} \alpha(S_{ij}) < \lambda(A \times B) + \varepsilon\,.$$

We may assume that $\{S_{ij}\}$ is exactly the set of all dyadic squares of side $1/2^N$ which intersect $A \times B$, and hence that

$$\bigcup S_{ij} = d_1 \times (e_1 \cup \cdots \cup e_m) \cup \cdots \cup d_n \times (e_1 \cup \cdots \cup e_m)\,,$$

where $\{d_1, \ldots, d_n\}$ is a non-overlapping covering of A and $\{e_1, \ldots, e_m\}$ is a non-overlapping covering of B. Thus

$$\begin{aligned} \lambda(A \times B) + \varepsilon &> \sum \alpha(S_{ij}) \\ &= \sum \ell(d_i) \sum \ell(e_j) \\ &\geq \mu(A)\,\mu(B)\,. \end{aligned}$$

The use of dyadic squares of the same size is necessary to ensure that we have a finite covering of $A \times B$ of the form $\{I_i \times J_j : i = 1, \ldots, n; \quad j = 1, \ldots, m\}$. ▮

Proposition 3. *If A and B are measurable subsets of $\mathbb{R}$, then*

$$\lambda(A \times B) = \mu(A)\,\mu(B)\,.$$

Proof. First assume that A and B have finite measure, so there are compact sets F and G, with $F \subset A$, and $G \subset B$, and

$$\mu(A) - \varepsilon < \mu(F),$$
$$\mu(B) - \varepsilon < \mu(G)\,.$$

Hence

$$\begin{aligned}\mu(A)\,\mu(B) &\geq \lambda(A \times B)\\ &\geq \lambda(F \times G)\\ &= \mu(F)\,\mu(G)\\ &> (\mu(A) - \varepsilon)(\mu(B) - \varepsilon)\,.\end{aligned}$$

Therefore the desired equality holds if A and B have finite measure.

If $\mu(A) = \infty$ or $\mu(B) = \infty$, let $A_n = A \cap [n, n+1), B_n = B \cap [n, n+1)$, for $n = 0, \pm 1, \pm 2, \ldots$. Then

$$A \times B = \bigcup_{n,m} A_n \times B_m\,,$$

the sets $A_n \times B_m$ are disjoint and measurable, and A_n, B_m have finite measure. Therefore,

$$\begin{aligned}\lambda(A \times B) &= \sum_{n,m} \lambda(A_n \times B_m)\\ &= \sum_{n,m} \mu(A_n)\mu(B_m)\\ &= \sum_n \mu(A_n) \sum_m \mu(B_m)\\ &= \mu(A)\mu(B)\,.\end{aligned}$$ ▮

Proposition 4. *If A and B are measurable subsets of $\mathbb{R}$, then $A \times B$ is a measurable subset of $\mathbb{R} \times \mathbb{R}$.*

Problem 2. Prove Proposition 4. Hint: Use Proposition 3, Chapter 9. If A and B are measurable, and $R = I \times J$ is a rectangle, then R is the disjoint union of $(A \cap I) \times (B \cap J)$, $(A' \cap I) \times (B \cap J)$, $(A \cap I) \times (B' \cap J)$, and $(A' \cap I) \times (B' \cap J)$. Moreover, we have $\mu(I) = \mu(A \cap I) + \mu(A' \cap I)$, $\mu(J) = \mu(B \cap J) + \mu(B' \cap J)$. ◂

In the general study of measure theory one starts with a σ-algebra of subsets of some set X, and a countably additive measure function ν on these specified "measurable sets." To define the product measure $\nu \times \nu$ on $X \times X$ one starts with the basic sets of the form $A \times B$, where A and B are measurable subsets of X. The product measure of such sets is of course $\nu{\times}\nu(A \times B) = \nu(A)\,\nu(B)$.

Our plane measure λ is of course the same as the product measure $\nu{\times}\nu$, although we did not define it that way. There is a fundamental intuitive idea of area, based on the area of a genuine rectangle, which is more or less independent of linear measure. We therefore used $\alpha(I \times J) = \ell(I)\,\ell(J)$ as our basic notion in defining λ on plane sets. This procedure was also designed to provide a second example of how an outer measure is defined in terms of some more basic notion. In the following proposition we show that λ could equally well have been defined starting with $\lambda(A \times B) = \mu(A)\,\mu(B)$ for *measurable* sets A and B rather than intervals.

Proposition 5. *For any set $E \subset \mathbb{R}^2$,*

$$\begin{aligned}\lambda(E) &= \inf\left\{\sum \lambda(A_i \times B_i) : E \subset \bigcup A_i \times B_i\right\}\\ &= \inf\left\{\sum \mu(A_i)\mu(B_i) : E \subset \bigcup A_i \times B_i\right\}\end{aligned}$$

where $\{A_i\}$ and $\{B_i\}$ are countable families of measurable subsets of $\mathbb{R}$.

Proof. Let $\lambda_m(E)$ denote the right side above; *i.e.*, $\lambda_m(E)$ would be the outer measure of E starting with measurable sets $A \times B$ rather than rectangles $I \times J$. Clearly $\lambda_m(E) \le \lambda(E)$ since the inf is over a larger collection of coverings. First suppose $E \subset [0,1) \times [0,1)$, so in particular $\lambda_m(E) < \infty$. Let $\{A_i \times B_i\}$ be a covering of E by measurable products, with $A_i \subset [0,1)$ and $B_i \subset [0,1)$, and

$$\sum_i \mu(A_i)\mu(B_i) < \lambda_m(E) + \varepsilon\,.$$

Let $\{I_{in}\}, \{J_{im}\}$ be coverings of A_i and B_i respectively by intervals, with

$$\begin{aligned}&\sum_n \ell(I_{in}) < \mu(A_i) + \varepsilon/2^i\,,\\ &\sum_m \ell(J_{im}) < \mu(B_i) + \varepsilon/2^i\,.\end{aligned}$$

Then the countable family $\{J_{in} \times J_{im}\}$ is a covering of E by rectangles, and so

$$\begin{aligned}
\lambda(E) &\leq \sum_i \sum_n \sum_m \ell(I_{in})\ell(J_{im}) \\
&= \sum_i \left(\sum_n \ell(I_{in})\right)\left(\sum_m \ell(J_{im})\right) \\
&< \sum_i \left(\mu(A_i) + \frac{\varepsilon}{2^i}\right)\left(\mu(B_i) + \frac{\varepsilon}{2^i}\right) \\
&\leq \sum_i \left(\mu(A_i)\mu(B_i) + \frac{\varepsilon}{2^i} + \frac{\varepsilon}{2^i} + \frac{\varepsilon^2}{4^i}\right) \\
&< \lambda_m(E) + 4\varepsilon .
\end{aligned}$$

Therefore $\lambda(E) \leq \lambda_m(E)$ if $E \subset [0,1) \times [0,1)$. The extension to general sets E is left as a problem. ▮

Problem 3. Finish the proof of Proposition 5 for arbitrary sets E. ◂

Now we have the following:

(i) If A and B are measurable subsets of $\mathbb{R}$, then $A \times B$ is a measurable subset of $\mathbb{R}^2$ and $\lambda(A \times B) = \mu(A)\,\mu(B)$.

(ii) If $E \subset \mathbb{R}^2$, then

$$\lambda(E) = \inf \left\{\sum \mu(A_i)\,\mu(B_i) : E \subset \bigcup (A_i \times B_i)\right\}$$

where the $\{A_i\}$ and $\{B_i\}$ are countable families of measurable subsets of $\mathbb{R}$.

We finish this chapter by showing how plane measure is related to linear measure through integration. In calculus one defines the area of the plane region

$$S = \{(x, y) : a \leq x \leq b,\ 0 \leq y \leq f(x)\}$$

to be the integral $\int_a^b f\,d\mu$. We now have a second definition of this area as $\lambda(S)$. It is necessary, and not hard, to show that these definitions agree.

Problem 4. If f is measurable and non-negative on $[a, b]$, and S is the region under the graph of f as given above, then S is measurable and $\lambda(S) = \int_a^b f\,d\mu$. ◂

We want to establish the connection between area and integration for more general plane regions, and thus pave the way for multiple integrals and the Fubini Theorem.

Let E be a measurable subset of the plane, and to start we will assume that $E \subset [0,1] \times [0,1]$. Let E_x be the cross-section of E over the point $x \in [0,1]$; *i.e.*,

$$E_x = \{y : (x, y) \in E\}.$$

Let $f(x) = \mu(E_x)$, so $f(x)$ is the "length" of the x-cross-section of E. We proceed to show that f is measurable and, as one would expect,

$$\lambda(E) = \int_0^1 f d\mu.$$

Problem 5 is an essential lemma for Proposition 6.

Problem 5. Let F be a compact subset of $[0,1] \times [0,1]$, and let F_x be the x-cross-section of F for each $x \in [0,1]$. For $x \in [0,1]$, let $L_n(x)$ be the total length $\sum \ell(e_i)$ of all dyadic intervals of length 2^{-n} which intersect F_x. Let $\alpha_n(F)$ be the total area $\sum \alpha(d_i \times e_j)$ of all dyadic squares of side 2^{-n} which intersect F.

(i) F is λ-measurable, and each F_x is μ-measurable.

(ii) $\alpha_n(F)$ decreases to $\lambda(F)$ as $n \longrightarrow \infty$.

(iii) $L_n(x)$ decreases to $\mu(F_x)$ as $n \longrightarrow \infty$.

(iv) What goes wrong if F is not compact?

Proposition 6. *If F is a compact subset of $[0,1] \times [0,1]$ and F_x is the x-cross-section, and $f(x) = \mu(F_x)$, then f is measurable and*

$$\lambda(F) = \int_0^1 f(x) d\mu(x).$$

Proof. Let $\{d_i \times e_j\}$ be all the dyadic squares of side 2^{-n} which intersect F. For x in the interior of d_i, let $\varphi_n(x)$ be the total length $\sum \ell(e_j)$ of all e_j such that $d_i \times e_j$ intersects F. That is, $\varphi_n(x)$ is the total length of the column of dyadic squares in the covering which lie over d_i. If x is an endpoint of d_i there may be two columns over x, so define $\varphi_n(x)$ only for interior points of the d_i, and let $\varphi_n(x) = 0$ elsewhere. Thus φ_n is a simple function and

$$\int \varphi_n d\mu = \sum \alpha(d_i \times e_j) = \alpha_n(F).$$

Moreover, $\varphi_n(x) \longrightarrow \mu(F_x)$ for all x which are not dyadic points; thus $\varphi_n(x) \longrightarrow \mu(F_x)$ a.e. It follows that $f(x) = \mu(F_x)$ is measurable, and

$$\lim_n \int_0^1 \varphi_n d\mu = \int_0^1 \mu(F_x)\, d\mu(x) = \int_0^1 f d\mu.$$

If n is so large that

$$\sum \alpha(d_i \times e_j) < \lambda(F) + \varepsilon$$

for the dyadic covering sets $d_i \times e_j$ of side 2^{-n}, then

$$\begin{aligned} \lambda(F) &\le \sum \alpha(d_i \times e_j) \\ &= \int_0^1 \varphi_n \, d\mu \\ &\le \lambda(F) + \varepsilon\,. \end{aligned}$$

Hence

$$\lambda(F) = \lim_n \int_0^1 \varphi_n d\mu = \int_0^1 \mu(F_x)\, d\mu(x). \quad ▮$$

Recall that λ is a regular measure, so for any measurable subset E of finite measure there is an open set U and a compact set F so that $F \subset E \subset U$ and

$$\lambda(E) - \varepsilon < \lambda(F) \le \lambda(E) \le \lambda(U) < \lambda(E) + \varepsilon\,.$$

If $E \subset (0,1) \times (0,1)$ then of course U and F can also be taken as subsets of $(0,1) \times (0,1)$.

Proposition 7. *If E is a measurable set, then $\mu(E_x)$ is measurable and $\int_0^1 \mu(E_x)\, d\mu(x) = \lambda(E)$.*

Proof. First assume that $E \subset (0,1) \times (0,1)$. Let $F_n \subset E \subset U_n$ with each F_n compact and each U_n open, and $\lambda(U_n - F_n) < \frac{1}{n}$.

Assume that $F_1 \subset F_2 \subset \cdots$ and $U_1 \supset U_2 \supset \cdots$. Let $K_n = [0,1] \times [0,1] - U_n$, so K_n is compact, and $\mu(F_{nx})$, $\mu(K_{nx})$ are measurable functions, with

$$\begin{aligned}
\lambda(F_n) &= \int_0^1 \mu(F_{nx})\,d\mu(x) \longrightarrow \lambda(E), \\
\lambda(U_n) &= 1 - \lambda(K_n) \\
&= 1 - \int_0^1 \mu(K_{nx})\,d\mu(x) \\
&= 1 - \int_0^1 (1 - \mu(U_{nx}))\,d\mu(x) \\
&= \int_0^1 \mu(U_{nx})\,d\mu(x) \longrightarrow \lambda(E).
\end{aligned}$$

If $f_n(x) = \mu(F_{nx})$, $g_n(x) = \mu(U_{nx})$, then $\{f_n\}$ and $\{g_n\}$ are monotone sequences of measurable functions with

$$f_n(x) \le \mu(E_x) \le g_n(x)$$

for all x, and

$$\lim_n \int_0^1 f_n d\mu = \lim_n \int_0^1 g_n d\mu = \lambda(E).$$

It follows that

$$\lim_n f_n(x) = \lim_n g_n(x) = \mu(E_x) \quad \text{a.e.,}$$

and so $\mu(E_x)$ is a measurable function and

$$\begin{aligned}
\int_0^1 \mu(E_x) d\mu &= \lim_n \int_0^1 f_n d\mu \\
&= \lambda(E). \quad \blacksquare
\end{aligned}$$

Problem 6. Extend Proposition 7 so that it applies to general measurable sets. ◀

Notice that the preceding results are proved first for subsets of the unit square. The final result is obtained by adding up what happens on a finite or countable number of squares. This is possible because both the line and the plane can be written as countable unions of finite measure sets. A general measure ν with this property is called σ-**finite**. Thus ν is a σ-finite measure on X if $X = \bigcup E_i$ with $\nu(E_i) < \infty$ for each i. The hypothesis of σ-finiteness is essential in the general discussion of product measures.

Proposition 7 provides the basis for the proof of the Fubini Theorem, which states that λ-integrals ("double integrals") can be evaluated as iterated μ-integrals. In practice, there is no other way to evaluate most double integrals. One of the most useful aspects of this connection between λ-integrals and μ-integrals is the fact that generally speaking the order of integration in iterated integrals can be reversed. The "Fubini Theorem" is the name commonly associated with the results of the next proposition, but the reader is cautioned that the name Tonelli is also given to this form.

Proposition 8. *(Fubini's Theorem) Let $f(x,y)$ be a non-negative measurable function, and define f_x by $f_x(y) = f(x,y)$. Then f_x is a measurable function on $\mathbb{R}$ for almost all x, and the function $F(x)$ defined by*

$$F(x) = \int_{\mathbb{R}} f_x(y)d\mu(y) = \int_{\mathbb{R}} f(x,y)d\mu(y)$$

is measurable and non-negative, and

$$\int_{\mathbb{R}} \left(\int_{\mathbb{R}} f(x,y)d\mu(y)\right) d\mu(x) = \int_{\mathbb{R}^2} f(x,y)d\lambda(x,y). \quad (1)$$

If f is integrable, f_x is integrable for almost all x and F is integrable; conversely, if the iterated integral is finite, then f is λ-integrable. Of course the same statements apply to the iteration in the other order, so the order of integration for iterated integrals (of a positive function) is immaterial.

Proof. First we give an outline of the proof.

For characteristic functions the result follows directly from Proposition 7. By linearity of the integrals – all three of them

– the result then holds for simple functions. A non-negative measurable function f is the limit of an increasing sequence φ_n of simple functions. This fact depends on the σ-finiteness of $\mathbb{R}^2$. (If $\mathbb{R}^2$ were not σ-finite we could not guarantee the existence of such functions φ_n which are zero off a set of finite measure; cf. Problem 8 below.) The theorem extends from simple functions to f by the Monotone Convergence Theorem.

Now we fill in the details. Let $f(x,y) = a\chi_E(x,y)$ where E is a measurable set and $\lambda(E) < \infty$. Then

$$f_x(y) = a\chi_{E_x}(y),$$

$$\int f_x(y)d\mu(y) = a\mu(E_x). \tag{2}$$

By Proposition 7,

$$\int a\mu(E_x)d\mu(x) = a\lambda(E). \tag{3}$$

Integrating both sides of (2), and using (3),

$$\int \left(\int f_x(y)d\mu(y) \right) d\mu(x) = \int a\mu(E_x)d\mu(x) = a\lambda(E);$$

i.e., for $f(x,y) = a\chi_E(x,y)$, we have the result:

$$\int \left(\int f(x,y)d\mu(y) \right) d\mu(x) = a\lambda(E) = \int f(x,y)d\lambda(x,y).$$

If $\varphi(x,y)$ is a simple function, then massive applications of linearity give the result. We illustrate the argument for the sum of two very simple functions. Let E_1 and E_2 be disjoint measurable sets of finite measure, and let $f = f_1 + f_2$ where $f_1 = a_1\chi_{E_1}$, $f_2 = a_2\chi_{E_2}$. Then

$$\begin{aligned}
\int f(x,y)d\lambda(x,y) &= \int (f_1 + f_2)d\lambda \\
&= \int f_1 d\lambda + \int f_2 d\lambda \\
&= \int \left(\int f_1(x,y)d\mu(y) \right) d\mu(x) \\
&\qquad + \int \left(\int f_2(x,y)d\mu(y) \right) d\mu(x) \\
&= \int \left(\int f_1(x,y)d\mu(y) + \int f_2(x,y)d\mu(y) \right) d\mu(x) \\
&= \int \left(\int (f_1(x,y) + f_2(x,y))d\mu(y) \right) d\mu(x).
\end{aligned}$$

Now let f be measurable and non-negative on $\mathbb{R}^2$. There is an increasing sequence $\{\varphi_n(x, y)\}$ of simple functions so that $\varphi_n \longrightarrow f$ a.e. By the Monotone Convergence Theorem

$$\int \varphi_n d\lambda \longrightarrow \int f d\lambda .$$

That is,

$$\int \left(\int \varphi_n(x, y) d\mu(y) \right) d\mu(x) \longrightarrow \int f d\lambda . \tag{4}$$

The right side of (4) could be $+\infty$.

For each fixed x, $\varphi_n(x, y)$ is an increasing sequence of simple functions, and

$$(\varphi_n)_x(y) = \varphi_n(x, y) \longrightarrow f_x(y) = f(x, y) .$$

Therefore,

$$\Phi_n(x) = \int \varphi_n(x, y) d\mu(y) \longrightarrow \int f(x, y) d\mu(y) = F(x) .$$

For each n,

$$\begin{aligned} \Phi_n(x) &= \int \varphi_n(x, y) d\mu(y) \\ &= \int \sum_{i=1}^{N} a_{ni} \chi_{E_{ni}}(x, y) d\mu(y) \\ &= \sum_{i=1}^{N} a_{ni} \mu(E_{ni})_x . \end{aligned}$$

By Proposition 7, it follows that each Φ_n is a measurable function of x, since $\mu(E_x)$ is measurable if E is. Since $\Phi_n(x)$ increases to $F(x)$, F is measurable. (F might take the value $+\infty$.) Hence the integral

$$\int f(x, y) d\mu(y)$$

is a measurable function of x. Finally, by the Monotone Convergence Theorem,

$$\begin{aligned} \int \Phi_n(x) d\mu(x) &= \int \left(\int \varphi_n(x, y) d\mu(y) \right) d\mu(x) \\ &\longrightarrow \int F(x) d\mu(x) \\ &= \int \left(\int f(x, y) d\mu(y) \right) d\mu(x) . \end{aligned}$$

By (4) we have the desired result for non-negative measurable functions f. The λ-integral is finite if and only if the iterated integral is finite.

If f is a not necessarily non-negative function, but is integrable, then the result holds for f^+ and f^- and hence for f. ▮

The following problem illustrates the way the Fubini Theorem is usually used to justify changing the order of integration in an iterated integral.

Problem 7. Let $f(x, y)$ be measurable. If

$$\iint |f(x,y)|\,d\mu(x)\,d\mu(y) < \infty$$

then $f(x, y)$ is λ-integrable and the order of integration can be reversed. ◀

Problem 8. Let X be an uncountable set, and $\mathcal{S}$ the σ-algebra of all subsets of X. Let μ be counting measure on X, so $\mu(E)$ is the number of elements in E if E is finite, and $\mu(E) = \infty$ otherwise. Show that the function f which is identically one on X is measurable, but there is no sequence $\{\varphi_n\}$ of simple functions which increases to f a.e. ◀

11 GENERAL MEASURES

The basic additivity property of Lebesgue linear or plane measure holds when the outer measure is restricted to the σ-algebra of measurable sets. We axiomatize the general theory of measure by starting with a σ-algebra $\mathcal{S}$ of subsets of some set X, and a countably additive non-negative extended real valued function ν defined on the sets in $\mathcal{S}$. This triple $(X, \mathcal{S}, \nu)$ is called a **measure space.** In practice, one says "ν is a measure on X." The σ-algebra $\mathcal{S}$ generally disappears from the discussion, and we write "E is measurable" to mean $\nu(E)$ makes sense - *i.e.*, that $E \in \mathcal{S}$.

Measures are automatically monotone in the sense that if A and B are measurable and $A \subset B$, then $\nu(A) \leq \nu(B)$. This is clear because $B - A$ is measurable and

$$\nu(B) = \nu(A) + \nu(B - A) \geq \nu(A)\,.$$

Countable subadditivity is also an automatic property of measures as we show next.

Proposition 1. *If $\{E_i\}$ is a finite or countable family of measurable sets, then*

$$\nu\left(\bigcup E_i\right) \leq \sum \nu(E_i)\,.$$

Proof. Let $F_n = E_n - (E_1 \cup \cdots \cup E_{n-1})$, so the F_n are measurable and disjoint, and $\bigcup F_n = \bigcup E_n$. Clearly $\nu(F_n) \leq \nu(E_n)$ for all n, so

$$\nu\left(\bigcup E_n\right) = \nu\left(\bigcup F_n\right) = \sum \nu(F_n) \leq \sum \nu(E_n)\,. \quad ▮$$

Problem 1. Let X be any set and for $E \subset X$, let $\nu(E)$ be the number of elements in E if E is a finite set, and $\nu(E) = \infty$ otherwise. Show that ν is a measure on the σ-algebra of all subsets of X. This measure is called *counting measure.*

Problem 2. Let X be an uncountable set. Let $\mathcal{S}$ be all subsets of X which are either countable or have a countable complement. Show that $\mathcal{S}$ is a σ-algebra. Let $\nu(E) = 0$ if E is countable and $\nu(E) = 1$ if E is uncountable. Show that ν is a measure on X. Hint: "Countable" means "countably infinite or finite."

Problem 3. Let f be a non-negative function on a set X. Let $\nu(E) = \sum_{x \in E} f(x)$, where the sum is the unordered sum in the sense of Problem 9, Chapter 2. Show that ν is a measure on all subsets of X. If X is uncountable and f is strictly positive, show that $\nu(X) = \infty$. [If f is the characteristic function of a single point x, then ν is called *point mass* at x, and $\int g\, d\nu = g(x)$ for all g.]

Proposition 2. *Let $\{E_n\}$ be a sequence of measurable sets.*
(i) If $E_1 \supset E_2 \supset \cdots$, and $\nu(E_1) < \infty$, then

$$\nu\left(\bigcap E_n\right) = \lim \nu(E_n).$$

(ii) If $E_1 \subset E_2 \subset \cdots$, then

$$\nu\left(\bigcup E_n\right) = \lim \nu(E_n).$$

Proof. Let $E_1 \supset E_2 \supset \cdots$ and let $E = \bigcap E_n$. Then

$$E_1 = (E_1 - E_2) \cup (E_2 - E_3) \cup \cdots \cup E,$$

and the summands are disjoint. Hence

$$\begin{aligned} \nu(E_1) &= (\nu(E_1) - \nu(E_2)) + (\nu(E_2) - \nu(E_3)) + \cdots + \\ &\quad + (\nu(E_{n-1}) - \nu(E_n)) + \cdots + \nu(E) \\ &= \nu(E_1) - \lim \nu(E_n) + \nu(E). \end{aligned}$$

Since $\nu(E_1) < \infty$, subtraction is legitimate and gives the result.

(iii) Let $E_1 \subset E_2 \subset \cdots$ and let $E = \bigcup E_n$. Then

$$E = E_1 \cup (E_2 - E_1) \cup (E_3 - E_2) \cup \cdots ,$$

and hence

$$\begin{aligned} \nu(E) &= \nu(E_1) + (\nu(E_2) - \nu(E_1)) + (\nu(E_3) - \nu(E_2)) + \cdots \\ &= \lim \nu(E_n). \end{aligned}$$

The two measures μ on $\mathbb{R}$ and λ on $\mathbb{R}\times\mathbb{R}$ already studied have the property that every set of (outer) measure zero is a measurable set, and consequently every subset of a set of measure zero is measurable. Measures with this property are called **complete**. In a general measure space $(X, \mathcal{S}, \nu)$ it is possible to have subsets A and B of X with $A \subset B \in \mathcal{S}$, and $\nu(B) = 0$, but $A \notin \mathcal{S}$, so $\nu(A)$ makes no sense. However, for any measure space $(X, \mathcal{S}, \nu)$ we can extend $\mathcal{S}$ to a larger σ-algebra $\mathcal{S}_0$ by throwing in all subsets of zero measure sets. The measure ν can then be extended to a complete measure ν_0 on $\mathcal{S}_0$ by defining $\nu_0(A) = \nu(B)$ if $B \in \mathcal{S}$ and A differs from B by a subset of a zero measure set; *i.e.*, if $A \Delta B \subset C$ with $\nu(C) = 0$.

Problem 4. Define a non-complete, non-trivial (*i.e.*, not identically zero) measure on a σ-algebra of subsets of the three-element set $X = \{a, b, c\}$.

Proposition 3. *Any measure can be extended to a complete measure.*

Proof. Let $(X, \mathcal{S}, \nu)$ be a measure space. For brevity let us say that a subset A of X is a *null-set* provided A is a subset of some $B \in \mathcal{S}$ with measure zero. Let $\mathcal{S}_0$ consist of all sets of the form $(E \cup A) - B$ where $E \in \mathcal{S}$ and A and B are null sets. Clearly $\mathcal{S} \subset \mathcal{S}_0$ since $\varnothing$ is a null set. We can assume that the sets in $\mathcal{S}_0$ have the form $(E \cup A) - B$ where A and B are null sets such that $E \cap A = \varnothing$ and $B \subset E$. In this case,

$$\begin{aligned} (E \cup A) - B &= (E - B) \cup A, \\ [(E \cup A) - B]' &= (E' \cup B) - A, \end{aligned}$$

so $\mathcal{S}_0$ is closed under complementation. Let $F_n = (E_n \cup A_n) - B_n \in \mathcal{S}_0$, with $E_n \in \mathcal{S}$, and A_n, B_n null sets. Then

$$\bigcup F_n = \bigcup [(E_n \cup A_n) - B_n] = \bigcup (E_n \cup A_n) - C\,,$$

where $C \subset \bigcup B_n$, and C is consequently a null set. Therefore

$$\bigcup F_n = \left(\bigcup E_n \cup \bigcup A_n\right) - C\,,$$

and $\bigcup A_n$, C are null sets. Hence $\mathcal{S}_0$ is a σ-algebra. We define

$$\nu_0[(E \cup A) - B] = \nu(E)\,.$$

To show that ν_0 is well defined, let

$$(E_1 \cup A_1) - B_1 = (E_2 \cup A_2) - B_2$$

with E_1, $E_2 \in \mathcal{S}$ and A_1, A_2, B_1, B_2 null sets. Let $\overline{A}_1$ be a zero measure set in $\mathcal{S}$ which contains A_1. Then

$$E_1 \cup A_1 \subset E_2 \cup A_2 \cup B_1\,,$$
$$E_1 \cup \overline{A}_1 \subset E_2 \cup A_2 \cup B_1 \cup \overline{A}_1\,.$$

Since $A_2 \cup B_1 \cup \overline{A}_1$ is a null set, there is $C \in \mathcal{S}$ with $\nu(C) = 0$ and $C \supset A_2 \cup B_1 \cup \overline{A}_1$. Hence $E_1 \cup \overline{A}_1 \subset E_2 \cup C$, and both of these sets are measurable. Hence

$$\nu(E_1) = \nu(E_1 \cup \overline{A}_1) \le \nu(E_2 \cup C) = \nu(E_2)\,.$$

The argument is symmetric, so $\nu(E_1) = \nu(E_2)$ and ν_0 is unambiguously defined. It is left as an exercise to show that ν_0 is countably additive on $\mathcal{S}_0$. ▮

Problem 5. (i) Show that every set in $\mathcal{S}_0$ can be written $F \cup C$ with $F \in \mathcal{S}$ and C a null set.

(ii) Show that ν_0 is complete and countably additive on $\mathcal{S}_0$. ◀

Problem 6. (i) What are $\mathcal{S}_0$ and ν_0 for the completion of your example in Problem 4?

(ii) Are the measures of Problems 1 and 2 complete? ◀

Since any measure can be extended to a complete measure, ν **is henceforth assumed to be a complete measure**. We will say that ν is **finite** if $\nu(X) < \infty$, and ν is σ-**finite** if X is a countable union of finite measure sets.

If ν_1 and ν_2 are two measures on the same σ-algebra of subsets of X, then it is clear that $\nu_1 + \nu_2$ is also a measure on X. More generally, any finite or countable sum $\sum a_i \nu_i$ with all $a_i > 0$ is again a measure on X if the ν_i are measures on the same σ-algebra. If we consider a difference $\nu_1 - \nu_2$ of two measures we again get a countably additive set function provided both measures are finite. However, if there is a set E such that $\nu_1(E) = \nu_2(E) = \infty$, then $(\nu_1 - \nu_2)(E)$ makes no sense. More complicated atrocities can happen. Since $\nu_1 - \nu_2$ could take both positive and negative values, it would be possible for $\sum_{i=1}^{\infty} (\nu_1 - \nu_2)(E_i)$ to converge conditionally for some disjoint sequence $\{E_i\}$. Conditional convergence means that the sum depends on the order of the summands. Since $\bigcup E_i$ obviously does not depend on the arrangement of the E_i, the countable additivity condition

$$(\nu_1 - \nu_2)\left(\bigcup E_i\right) = \sum (\nu_1 - \nu_2)(E_i)$$

would be an impossibility in this case.

Problem 7. Show that if ν_1 and ν_2 are finite measures on the same σ-algebra of subsets of X, then $\nu_1 - \nu_2$ is countably additive on this σ-algebra.

Now let us consider general countably additive set functions ν which take both positive and negative values. We will call such a function a **signed measure**. The countable additivity condition requires that all sums $\sum \nu(E_i)$ be absolutely convergent when $\{E_i\}$ is a disjoint family of measurable sets. We know from Problem 7 that some signed measures arise as the difference of two positive measures, and we will show below that every signed measure is such a difference of two positive measures.

Let ν be a signed measure on X. We will allow some sets to have measure $+\infty$, but no sets to have measure $-\infty$. The other choice of allowing $-\infty$ but not $+\infty$ would work just as well. A measurable set A is called a **positive set** provided $\nu(E) \geq 0$ for every measurable subset E of A, including A itself. A measurable set B is a **negative set** provided $\nu(E) \leq 0$ for all measurable $E \subset B$. If A is both positive and negative, so that every subset of A has measure zero, then A is called a **null set**. Notice that "null set" in this context is different from "null set" in the earlier completion argument.

Problem 8. If ν is a signed measure on X and A is a positive set for ν, and $\nu_1(E) = \nu(E \cap A)$, then ν_1 is a positive measure on X. If B is a negative set for ν and $\nu_2(E) = -\nu(E \cap B)$, then ν_2 is a positive measure on X. ◀

We proceed to show that every signed measure ν can be written as the difference of positive measures as indicated in the preceding problem, where B is a "largest" negative set for ν and $A = X - B$. The problem is to show that there is a "largest" negative set B, so that $X - B$ is necessarily a positive set. We find B first rather than A because of the possible complications stemming from the fact that $\nu(A)$ might be $+\infty$.

Proposition 4. *Let ν be a signed measure on X. If $\nu(E) < 0$ then E contains a negative set of negative measure. If $\nu(E) > 0$ then E contains a positive set of positive measure.*

Proof. We will prove the first statement so let E be a measurable set with $\nu(E) < 0$. If there are no subsets of E with positive measure, then E is a negative set and we are done. Otherwise, let

$$p_1 = \sup\{\nu(F) : F \subset E\} > 0.$$

If there is $F_1 \subset E$ with $\nu(F_1) = p_1$, then clearly $E - F_1$ is a negative set of negative measure and we are done. Otherwise, pick $F_1 \subset E$ so

$$0 < p_1/2 < \nu(F_1) < p_1.$$

Let

$$p_2 = \sup\{\nu(F) : F \subset E - F_1\}$$

and notice that $0 < p_2 < p_1/2$. If there is $F_2 \subset E - F_1$ so that $\nu(F_2) = p_2$, then $E - (F_1 \cup F_2)$ is a negative set of negative measure. Otherwise, we pick $F_2 \subset E - F_1$ with

$$0 < p_2/2 < \nu(F_2) < p_2 < p_1/2.$$

Continuing this way we either arrive at a negative set $E - (F_1 \cup \cdots \cup F_n)$ after n steps, or we define a disjoint sequence $\{F_n\}$ of subsets of E such that

$$0 < p_n/2 < \nu(F_n) < p_n < p_1/2^{n-1},$$

where

$$p_n = \sup\{\nu(F) : F \subset E - (F_1 \cup \cdots \cup F_{n-1})\}.$$

We show that $E - \bigcup F_n$ is a negative set, and of course $\nu(E - \bigcup F_n) < \nu(E) < 0$. If there is $F \subset E - \bigcup F_n$ with $\nu(F) > 0$, then $p_n \geq \nu(F)$ for all n, which contradicts $p_n < p_1/2^{n-1} \longrightarrow 0$. ▮

Proposition 5. *If ν is a signed measure on X, then there is a positive set A and a negative set B so that A and B are disjoint and $A \cup B = X$.*

Proof. If there are no measurable sets of negative measure, then we take $B = \emptyset$, $A = X$, and we are done. Otherwise, there are sets of negative measure, and hence negative sets of negative measure. Let

$$q = \inf\{\nu(E) : E \text{ a negative set}\}.$$

Let $\{B_n\}$ be a sequence of negative sets such that $\nu(B_n) \longrightarrow q < 0$. The union of negative sets is a negative set, so we can assume that $B_1 \subset B_2 \subset B_3 \subset \cdots$. Let $B = \bigcup B_n$, so that B is a negative set and

$$\nu(B) = \lim \nu(B_n) = q.$$

Clearly there is no set of negative measure disjoint from B, so $A = X - B$ is a positive set. ▮

A decomposition of X into disjoint sets A and B, with A a positive set and B a negative set, is called a **Hahn decomposition** of X. Such a decomposition is not unique since any null set can be thrown into either A or B. However, apart from null sets, "the" Hahn decomposition is unique.

Problem 9. Let ν be a signed measure on X and let A_1 and A_2 be positive sets for X and B_1, B_2 negative sets for X, with

$X = A_1 \cup B_1 = A_2 \cup B_2$ and $A_1 \cap B_1 = A_2 \cap B_2 = \varnothing$. Show that $A_1 \Delta A_2$ is a null set, which proves that the Hahn decomposition is unique except for null sets. ◀

If A and B form a Hahn decomposition of X for the signed measure ν, then we know from Problem 8 that ν can be written as the difference of two measures, $\nu = \nu^+ - \nu^-$, where

$$\nu^+(E) = \nu(E \cap A), \quad \nu^-(E) = \nu(E \cap B).$$

The two measures ν^+ and ν^- are supported on the disjoint sets A and B in the sense that $\nu^+(B) = 0$ and $\nu^-(A) = 0$. We will say that two (positive) measures ν_1 and ν_2 on X are **singular** (or **mutually singular**, or **singular with respect to each other**), provided there are disjoint sets A and B with $A \cup B = X$ and $\nu_1(B) = \nu_2(A) = 0$. The canonical representation $\nu = \nu^+ - \nu^-$ of a signed measure ν as the difference of two singular measures is called the **Jordan decomposition** of ν. The Hahn decomposition of X into a positive set A and a negative set B is only unique up to null sets. However, the null sets do not affect the Jordan decomposition $\nu = \nu^+ - \nu^-$ of ν into singular measures, because null sets do not affect the definition of ν^+ and ν^-. Therefore the Jordan decomposition $\nu = \nu^+ - \nu^-$ is unique.

If ν is a signed measure, then we define the **absolute value** of ν or **total variation** of ν by

$$|\nu|(E) = \nu^+(E) + \nu^-(E).$$

Observe that $|\nu|$ is again a measure on X, and $|\nu(E)| \leq |\nu|(E)$ for all E.

12 INTEGRATION FOR GENERAL MEASURES

In this chapter we let $(X, \mathcal{S}, \nu)$ be a general measure space and define the integral with respect to ν. Measures are non–negative unless otherwise specified, and all measures are complete. We consider first bounded functions defined on sets of finite measure, and measurability is again the critical property. For convenience, the definitions and theorems which are concerned solely with measurability will be understood to include both bounded and unbounded functions and extended real valued functions, since these facts will be needed later. The reader is encouraged to notice that our development here is basically the same as that for integrals on the line (Chapter 5), and to consider specifically what each statement means for integrals with respect to plane measure λ.

Let f be a bounded function defined on a measurable set S of finite measure. A partition $P = \{E_1, \ldots, E_n\}$ of S is again a finite collection of disjoint measurable sets E_i whose union is S. Another partition $Q = \{F_1, \ldots, F_m\}$ is a refinement of P, denoted $Q \succ P$ or $P \prec Q$, provided each F_i is a subset of some E_j. The partitions of S are partially ordered by refinement and so form a directed set. The integral of a primary function will be the limit of several nets defined on this directed set.

If P is a partition of S, the upper and lower sums for f and P are defined as usual:

$$L(f,P) = \sum m_i \nu(E_i),$$
$$U(f,P) = \sum M_i \nu(E_i),$$

where m_i and M_i denote the inf and sup of the values $f(x)$ for $x \in E_i$. The lower sums $\{L(f,P)\}$ and upper sums $\{U(f,P)\}$

are nets on the set of all partitions of S. As before, lower sums increase and upper sums decrease as the partition is refined, and all lower sums are less than all upper sums. These monotone nets converge and

$$\begin{aligned}\lim_P L(f,P) &= \sup_P L(f,P)\\ &\le \inf_P U(f,P) = \lim_P U(f,P).\end{aligned}$$

If the limits are equal we say that f is **integrable** over S, and write $\int_S f\,d\nu$ for the common limit.

If $P = \{E_i\}$ is a partition of S and $c_i \in E_i$ for each i, then the corresponding Riemann sum is

$$R(f,P,c) = \sum f(c_i)\nu(E_i).$$

The Riemann sums $R(f,P,c)$ form a net on all pairs (P,c), which we partially order with the usual refinement order on partitions; *i.e.*,

$$(P,c) \succ (P',c') \quad \text{if} \quad P \succ P'.$$

For any choice c for a partition P,

$$L(f,P) \le R(f,P,c) \le U(f,P).$$

It follows that if f is integrable over S,

$$\int_S f\,d\nu = \lim_P R(f,P,c),$$

where we write "$\lim_P$" for convenience rather than "$\lim_{(P,c)}$" since the ordering depends only on P.

Problem 1. If f is bounded on a measurable set of finite measure, and $\lim_P R(f,P,c)$ exists, then f is integrable (and the integral is therefore the limit).

Problem 2. If a bounded function f is integrable over a set S of finite measure, and T is a measurable subset of S, then f is integrable over T and

$$\int_T f\,d\nu = \int_S f\chi_T\,d\nu.$$

Problem 3. If a bounded function f is integrable over a set S of finite measure, and g is a bounded function which equals f a.e. on S, then g is integrable and $\int_S g\,d\nu = \int_S f\,d\nu$.

Proposition 1. *If f and g are bounded integrable functions on the finite measure set S, and a is a constant, then af and $f+g$ are integrable and*

$$\int_S af\,d\nu = a\int_S f\,d\nu,$$
$$\int_S (f+g)\,d\nu = \int_S f\,d\nu + \int_S g\,d\nu.$$

Proof. For any partition $P = \{E_i\}$ and any choice function c for P (so $c_i \in E_i$) we have

$$R(af,P,c) = aR(f,P,c),$$
$$R(f+g,P,c) = R(f,P,c) + R(g,P,c).$$

The convergence of the nets $\{R(f,P,c)\}$ and $\{R(g,P,c)\}$ implies the convergence of the nets on the right above, and hence

$$\int_S af\,d\nu = \lim_P R(af,P,c) = a\lim_P R(f,P,c) = a\int_S f\,d\nu,$$
$$\begin{aligned}\int_S (f+g)\,d\nu &= \lim_P R(f+g,P,c)\\ &= \lim_P R(f,P,c) + \lim_P R(g,P,c)\\ &= \int_S f\,d\nu + \int_S g\,d\nu. \quad \blacksquare\end{aligned}$$

The properties of the integral listed in the next problem are also immediate consequences of properties of nets and the fact that if f is bounded and integrable on a set of finite measure, the integral is a limit of any one of the three nets $\{L(f,P)\}$, $\{U(f,P)\}$, $\{R(f,P,c)\}$.

Problem 4. Let S be a set of finite measure, and let f and g be bounded and integrable over S. Let A and B be disjoint measurable subsets of S.

(i) If $f \geq 0$ on S, then $\int_S f\,d\nu \geq 0$.

(ii) If $f \geq g$ on S, then $\int_S f\,d\nu \geq \int_S g\,d\nu$.

(iii) $|f|$ is integrable on S, and $\int_S |f|\,d\nu \geq \left|\int_S f\,d\nu\right|$.

(iv) $\int_{A\cup B} f\,d\nu = \int_A f\,d\nu + \int_B f\,d\nu$. ◀

Problem 5. If φ is a simple function defined on S, with $\varphi = \sum_{i=1}^{N} a_i \chi_{E_i}$ where the E_i are measurable but not necessarily disjoint, then φ is integrable and

$$\int_S \varphi\,d\nu = \sum_{i=1}^{N} a_i \nu(E_i). \quad ◀$$

If $P = \{E_i\}$ is a partition of S and m_i, M_i are the infs and sups of f on the E_i, then

$$\varphi_P = \sum m_i \chi_{E_i}, \quad \psi_P = \sum M_i \chi_{E_i},$$

are simple functions with

$$\varphi_P \le f \le \psi_P,$$

and

$$\int \varphi_P\,d\nu = \sum m_i \nu(E_i) = L(f,P),$$
$$\int \psi_P\,d\nu = \sum M_i \nu(E_i) = U(f,P).$$

Hence if a bounded function f is integrable on a set of finite measure there are simple functions φ and ψ with $\varphi \le f \le \psi$ and $\int \psi\,d\nu - \int \varphi\,d\nu < \varepsilon$. Conversely, suppose there are such simple functions $\varphi = \sum c_i \chi_{E_i}$, $\psi = \sum d_j \chi_{F_j}$ with $\varphi \le f \le \psi$ and $\int \psi\,d\nu - \int \varphi\,d\nu < \varepsilon$. We may assume that $\{E_i\}$ and $\{F_j\}$ are partitions of S. Let P be the partition consisting of all $E_i \cap F_j$. Then

$$\int \varphi\,d\nu \le L(f,P) \le U(f,P) \le \int \psi\,d\nu,$$

so $U(f,P) - L(f,P) < \varepsilon$ and f is integrable. We have proved the following proposition.

Proposition 2. *If f is a bounded function on a set S of finite measure, then f is integrable if and only if there are simple functions φ and ψ with $\varphi \leq f \leq \psi$ and $\int \psi d\nu - \int \varphi d\nu < \varepsilon$.*

Most texts define "f is integrable" to mean the existence of simple functions φ and ψ as in Proposition 2. We retain the upper-sum, lower-sum definition to avoid the appearance of an artificial difference in the way the Riemann and Lebesgue integrals are defined. The real difference lies in the existence of a measure for the partitioning subsets $\{E_i\}$.

The fact that an integrable function is one which can be squeezed between simple functions provides the characterization of integrable functions as measurable functions. First we show that a bounded integrable function is the pointwise limit (a.e.) of simple functions, and then show that such limits can be characterized by the usual definition of measurability. The procedure here is slightly different from that of Chapter 5 in that here we first "discover" that measurability is necessary for integrability, and then notice that measurability also suffices for integrability of bounded functions on sets of finite measure. This difference in approach is purely for the sake of variety. We will need the result of the following problem.

Problem 6. If $\{\varphi_n\}$ is a decreasing sequence of non-negative simple functions on a set S of finite measure, and $\int_S \varphi_n d\nu \longrightarrow 0$, then $\varphi_n \longrightarrow 0$ a.e. Show that the almost everywhere restriction is a necessary one. ◀

Now suppose that f is a bounded function which is integrable over S, and let $\{P_n\}$ be a sequence of partitions of S with $P_1 \prec P_2 \prec P_3 \prec \cdots$, so that $U(f, P_n)$ decreases to $\int f d\nu$ and $L(f, P_n)$ increases to $\int f d\nu$. Let φ_n and ψ_n be the simple functions corresponding to the lower and upper sums, so that $\varphi_n \leq f \leq \psi_n$, $\{\varphi_n\}$ is increasing, and $\{\psi_n\}$ is decreasing. Then $\{\psi_n - \varphi_n\}$ is a decreasing sequence of non-negative simple functions with

$$\int_S (\psi_n - \varphi_n) d\nu \longrightarrow 0.$$

It follows that $\psi_n - \varphi_n \longrightarrow 0$ a.e., and hence,

$$\varphi_n \longrightarrow f, \quad \psi_n \longrightarrow f \quad \text{a.e.}$$

We have proved the following:

Proposition 3. *If f is a bounded function which is bounded on set S of finite measure, then there are increasing simple functions $\{\varphi_n\}$ and decreasing simple functions $\{\psi_n\}$ with $\varphi_n \leq f \leq \psi_n$ for all n and*

$$f = \lim \varphi_n = \lim \psi_n \quad \text{a.e.}$$

Proposition 4. *If f is any bounded function on a set S of finite measure, then f is the a.e. limit of an increasing sequence of simple functions if and only if $\{x : a < f(x)\}$ is measurable for all a.*

Proof. Assume first that $\varphi_n(x)$ increases to $f(x)$ for all x. Then $a < f(x)$ if and only if $a < \varphi_n(x)$ for some n. Hence

$$\{x : a < f(x)\} = \bigcup_n \{x : a < \varphi_n(x)\}.$$

Each set on the right is a finite union of measurable sets since φ_n is simple, so $\{x : a < f(x)\}$ is measurable. If $\varphi_n \longrightarrow f$ a.e., then $\{x : a < f(x)\}$ differs from the measurable set $\{x : a < \lim \varphi_n(x)\}$ by a set of measure zero, and is therefore also measurable.

Now assume that $\{x : a < f(x)\}$ is measurable for all a. Then

$$\{x : a \leq f(x)\} = \bigcap_n \left\{x : a - \frac{1}{n} < f(x)\right\}$$

is measurable for all a, as is

$$\{x : a \leq f(x) < b\} = \{x : a \leq f(x)\} - \{x : b \leq f(x)\}.$$

Let

$$\varphi_n = \frac{k}{2^n} \quad \text{on} \quad \left\{x : \frac{k}{2^n} \leq f(x) < \frac{k+1}{2^n}\right\}$$

whenever this set is non-empty, and $\varphi_n = 0$ elsewhere. Then φ_n is a simple function and $\{\varphi_n(x)\}$ increases to $f(x)$ for all $x \in S$. ▮

We define f to be **measurable** provided $\{x : a \leq f(x) < b\}$ is measurable for all a and b. As before, the following are equivalent conditions:

$$\{x : f(x) > a\} \text{ is measurable for all } a;$$
$$\{x : f(x) < b\} \text{ is measurable for all } b;$$
$$\{x : f(x) \leq b\} \text{ is measurable for all } b;$$
$$\{x : f(x) \geq a\} \text{ is measurable for all } a;$$
$$\{x : a \leq f(x) < b\} \text{ is measurable for all } a, b.$$

If f is an extended real valued function then we require that $\{x : f(x) = \infty\}$ and $\{x : f(x) = -\infty\}$ be measurable.

Problem 7. (i) Show that $f + g$ is measurable if f and g are.

(ii) Show that fg is measurable if f and g are. Hint: $fg = \frac{1}{4}[(f+g)^2 - (f-g)^2]$. ◀

The following characterization of integrability is now in hand; the "integrable implies measurable" part is proved above, and the "measurable implies integrable" part is proved exactly as in Chapter 5.

Proposition 5. *If f is a bounded function on a set of finite measure, then f is integrable if and only if f is measurable.*

As we saw in the earlier chapters, the utility of the above characterization stems from the fact that limits of measurable functions are measurable.

Proposition 6. *If $\{f_n\}$ is a sequence of measurable functions, then $\sup f_n$, $\inf f_n$, $\limsup f_n$, $\liminf f_n$, are all measurable; $\lim f_n$ is measurable if the limit exists. Simple functions are measurable.*

Problem 8. Show that $\sup f_n$ is measurable if each f_n is measurable. ◀

Egoroff's Theorem remains intact in our general setting. Thus if $\{f_n\}$ is a sequence of measurable functions on a set S of finite measure, and $f_n \longrightarrow f$ on S with all functions being finite valued but not necessarily bounded, then $f_n \longrightarrow f$ uniformly off sets of arbitrarily small measure. It follows that $\int_S f_n d\nu \longrightarrow \int_S f d\nu$ in case the f_n are uniformly bounded.

Proposition 7. *(Egoroff's Theorem). If f_n is measurable on S for each n and and $\nu(S) < \infty$, and $f_n \longrightarrow f$ a.e. on S, then given $\delta > 0$ there is a set E with $\nu(E) < \delta$ so that $f_n \longrightarrow f$ uniformly on $S - E$.*

Proof. (Outline) Assume without loss that $f_n(x) \longrightarrow f(x)$ for all $x \in S$. For $\varepsilon > 0$ define

$$A_N(\varepsilon) = \{x \in S : |f_k(x) - f(x)| \geq \varepsilon \text{ for some } k \geq N\} .$$

Then $\bigcap_N A_N(\varepsilon) = \varnothing$ and $\nu\left(A_N(\varepsilon)\right) \longrightarrow 0$ for all $\varepsilon > 0$. Let $\delta > 0$. Pick N_1, so that $\nu\left(A_{N_1}(1)\right) < \delta/2$ and N_2 so that $\nu\left(A_{N_2}(\frac{1}{2})\right) < \delta/4$ and so on: $\nu\left(A_{N_n}(\frac{1}{n})\right) < \delta/2^n$. Let $E = \bigcup_n A_{N_n}(\frac{1}{n})$, so $\nu(E) < \delta$, and $f_n \longrightarrow f$ uniformly off E. ▥

Problem 9. Fill in the details of the proof of Egoroff's Theorem. ◀

Problem 10. If $\{f_n\}$ is a sequence of measurable functions on a set S of finite measure, and $|f_n(x)| \leq M$ for all n and all $x \in S$, and $f_n \longrightarrow f$ a.e., then $\int_S f_n d\nu \longrightarrow \int_S f d\nu$. (This is the Bounded Convergence Theorem again.) ◀

13 MORE INTEGRATION--
THE RADON--NIKODYM THEOREM

In this chapter we will complete the definition of $\int_X f d\nu$, where ν is a measure on X, and we now admit the possibility that $\nu(X) = \infty$ and that f is unbounded. We also now consider extended real valued functions, so that $f(x) = +\infty$ and $f(x) = -\infty$ are possibilities.

We will again call the bounded measurable functions defined (or non-zero) only on sets of finite measure the **primary functions**. We also agree that if a function f is defined only on a measurable set S we will, when convenient, consider it to be defined on all of X by letting $f \equiv 0$ on $X - S$. The extended f is of course also measurable. The primary functions are exactly the functions for which $\int_X f d\nu$ has already been defined.

Integrability over infinite measure sets is again defined to be absolute integrability, so we consider first **non-negative measurable functions** f on X, and define

$$\int f d\nu = \sup \left\{ \int g d\nu : 0 \leq g \leq f,\ g \text{ primary} \right\} .$$

If f is any non-negative function - measurable or not, bounded or not - the above definition makes sense. For example, let f be the characteristic function of the non-measurable subset E of $[0, 1]$ constructed in Chapter 4. Such a set E has no measurable subsets of positive measure (Problem 1) so $\int g d\mu = 0$ for all primary functions g under f, and hence the above definition would give us the "definition" $\int f d\mu = 0$. We do not want such a definition because we have already decided that this kind of function is not worthy of integrability. Therefore, **we henceforth consider only measurable functions on** X.

Problem 1. Show that the non-measurable subset E of $[0,1]$ constructed in Chapter 4 has no measurable subsets of positive measure. Verify that $\int g\,d\mu = 0$ for all measurable g such that $0 \le g \le \chi_E$.

For any non-negative measurable f we have a value for $\int f\,d\nu$, but that value may be $+\infty$. We say that f is **integrable** provided f is measurable and the integral is finite.

Problem 2. The improper Riemann integral of $1/\sqrt{x}$ over $[0,1]$ is 2. Show the Lebesgue integral is also 2.

Problem 3. If f is a non-negative function on $[0,1]$ which is continuous on $(0,1]$, show that the Riemann and Lebesgue integrals of f over $[0,1]$ are the same, including the case where both are $+\infty$.

Problem 4. (i) Let X be an infinite set and let $\varnothing$ and X be the only measurable subsets of X, with $\nu(\varnothing) = 0$, $\nu(X) = \infty$. What are the measurable functions? What is $\int 1\,d\nu$?

(ii) Now suppose that X is an infinite set with $\nu(X) = \infty$, but there is a countable family $\{X_n\}$ of subsets of X with $\nu(X_n) < \infty$ for all n and $X = \bigcup X_n$; that is, suppose X is σ-finite. Now what is $\int 1\,d\nu$?

Proposition 1. *If f and g are non-negative measurable functions, and $a > 0$, then af and $f + g$ are measurable, and*

(i) $\int af\,d\nu = a\int f\,d\nu$,
(ii) $\int (f+g)\,d\nu = \int f\,d\nu + \int g\,d\nu$.

Proof. The proof of (i) is immediate from the definition since ag is a primary function under af if and only if g is a primary function under f. To prove (ii), let f_1 and g_1 be primary functions with $0 \le f_1 \le f$ and $0 \le g_1 \le g$. Then $f_1 + g_1$ is a primary function and $0 \le f_1 + g_1 \le f + g$, so

$$\int (f+g)\,d\nu \ge \int (f_1 + g_1)\,d\nu = \int f_1\,d\nu + \int g_1\,d\nu .$$

This holds for all $f_1 \le f$ and $g_1 \le g$, so

$$\int (f+g)\,d\nu \ge \int f\,d\nu + \int g\,d\nu .$$

Now let h be any primary function with $0 \le h \le f + g$. Let $h_1 = h \wedge f$ and $h_2 = h - h_1$, so h_1 and h_2 are primary functions (both bounded above by the bound on h, and non-zero only where h is non-zero). Since $h_1 \le f$ and $h_2 \le g$,

$$\int h d\nu = \int h_1 d\nu + \int h_2 d\nu \le \int f d\nu + \int g d\nu ,$$

and this holds for all primary h under $f + g$, so

$$\int (f + g) d\nu \le \int f d\nu + \int g d\nu .$$

Notice that f and g are not required to be integrable, so $f + g$ is integrable if and only if both f and g are. ▮

The basic convergence theorem for integrals is again Fatou's Lemma.

Proposition 2. *(Fatou's Lemma). If $\{f_n\}$ is a sequence of non-negative measurable functions, and $f_n \longrightarrow f$ a.e., then*

$$\liminf \int f_n d\nu \ge \int f d\nu .$$

If f is not integrable, both sides equal $+\infty$.

Proof. Let h be a non-negative primary function with $h \le f$. Let $h_n = h \wedge f_n$ so that $\{h_n\}$ is a uniformly bounded sequence of primary functions, and all h_n are non-zero only on the set where h is non-zero. Since $f_n \longrightarrow f \ge h$, $h_n \longrightarrow h$ and consequently, by the bounded convergence theorem,

$$\int h_n d\nu \longrightarrow \int h d\nu .$$

Since $f_n \ge h_n$,

$$\liminf \int f_n d\nu \ge \lim \int h_n d\nu = \int h d\nu .$$

This inequality holds for all primary $h \le f$, so

$$\liminf \int f_n d\nu \ge \int f d\nu . \quad ▮$$

Corollary. *(Monotone Convergence Theorem) If $\{f_n\}$ is a sequence of non-negative measurable functions such that $f_n \le f$ for all n and $f_n \longrightarrow f$, then $\int f_n d\nu \longrightarrow \int f d\nu$. In particular, if $\{f_n\}$ increases to f, then $\int f_n d\nu \longrightarrow \int f d\nu$.*

Problem 5. (i) The proof above of Proposition 2 tacitly assumes that $f_n \longrightarrow f$ everywhere. Supply the details for the case $f_n \longrightarrow f$ a.e.

(ii) Check through the proof of Fatou's Lemma in case $\int f d\nu = \infty$. Is anything else needed?

Problem 6. If $\{f_n\}$ is a sequence of non-negative measurable functions, then $\lim f_n$ may not exist, but $\liminf f_n$ and $\limsup f_n$ are measurable functions. What can you say about relations between the following four numbers?

$$\liminf \int f_n d\nu , \quad \limsup \int f_n d\nu ,$$
$$\int \liminf f_n d\nu , \quad \int \limsup f_n d\nu .$$

Hint: Problem 17(ii), Chapter 7, provides a relevant example.

Problem 7. Let $\{f_n\}$ be a sequence of non-negative measurable functions. Show that $\sum f_n$ is measurable and

$$\int \sum f_n d\nu = \sum \int f_n d\nu .$$

For an arbitrary measurable function f we again write $f = f^+ - f^-$ where $f^+ = f \vee 0$ and $f^- = (-f) \vee 0$. We say f is **integrable** provided both f^+ and f^- are integrable, in which case we define

$$\int f d\nu = \int f^+ d\nu - \int f^- d\nu .$$

The additivity properties of the integral are proved just as before, and the proof of the following proposition is left as an exercise.

Proposition 3. *If f and g are integrable and a is a constant, then af and $f + g$ are integrable, and*

(i) $\int af d\nu = a \int f d\nu$,

(ii) $\int (f+g)d\nu = \int f d\nu + \int g d\nu$,
(iii) $|f|$ *is integrable and* $|\int f d\nu| \le \int |f| d\nu$,
(iv) *If* $g \le f$, $\int g d\nu \le \int f d\nu$.

Problem 8. Prove Proposition 3.

Proposition 4. *If f is integrable and E is a measurable set, then $f\chi_E$ is integrable.*

Proof. This is clear if $f \ge 0$, since if h is primary and $0 \le h \le f\chi_E$, then $0 \le h \le f$ and $\int f\chi_E d\nu \le \int f d\nu$. If $f = f^+ - f^-$ then $f\chi_E = f^+\chi_E - f^-\chi_E$ so $f\chi_E$ is integrable.

We define, for measurable subsets E of X,

$$\int_E f d\nu = \int f\chi_E d\nu .$$

Problem 9. (i) If $\{E_n\}$ is a disjoint sequence of measurable sets, and $E = \bigcup E_n$, and f is integrable, then

$$\int_E f d\nu = \sum \int_{E_n} f d\nu .$$

(ii) If f is integrable and we define β on measurable subsets of X by

$$\beta(E) = \int_E f d\nu ,$$

then β is a signed measure on X and $|\beta(E)| < \infty$ for all E. If $\nu(E) = 0$, then $\beta(E) = 0$.

Fatou's Lemma and the Monotone Convergence Theorem show that pointwise convergence of functions, $f_n \longrightarrow f$, implies that the integrals converge unless the functions f_n are allowed to reach out into an infinite area outside the graph of f. If we demand that all f_n lie in some finite area, then $f_n \longrightarrow f$ implies $\int f_n \longrightarrow \int f$.

Proposition 5. *(Lebesgue Dominated Convergence Theorem). If g is integrable and $\{f_n\}$ is a sequence of measurable functions such that $|f_n| \le g$ for all n, and $f_n \longrightarrow f$ a.e., then*

$$\int f_n d\nu \longrightarrow \int f d\nu .$$

Problem 10. Prove the Lebesgue Dominated Convergence Theorem. Hint: Consider the non-negative sequences $g + f_n \longrightarrow g + f$ and $g - f_n \longrightarrow g - f$, and apply Fatou's Lemma. ◀

Recall that if f is a bounded integrable function on the line, and

$$F(x) = \int_0^x f d\mu ,$$

then F is uniformly continuous. This is easy to see, since if $|f(x)| \leq M$,

$$|F(x_1) - F(x_2)| = \left| \int_{[x_1,x_2]} f d\mu \right| \leq M |x_2 - x_1| .$$

As a consequence, if $\varepsilon > 0$ there is $\delta > 0$ (*e.g.* $\delta = \varepsilon / M$) so that whenever $\{(a_n, b_n)\}$ is a sequence of disjoint open intervals on the line such that $\sum(b_n - a_n) < \delta$, then $\sum |F(b_n) - F(a_n)| < \varepsilon$. This will also be an immediate consequence of the next proposition, which can be stated informally thus: for all integrable functions, bounded or not,

$$\int_E f d\nu \longrightarrow 0 \text{ as } \nu(E) \longrightarrow 0 .$$

Proposition 6. *If f is integrable over X and $\varepsilon > 0$, there is $\delta > 0$ so that $|\int_E f d\nu| < \varepsilon$ whenever $\nu(E) < \delta$.*

Proof. It is sufficient to prove this for non-negative f (Problem 10). So assume f is a non-negative integrable function, and let h be a primary function with $0 \leq h \leq f$ and

$$\int f d\nu - \int h d\nu < \varepsilon .$$

For any measurable set E we obviously have

$$\int_E f d\nu < \int_E h d\nu + \varepsilon .$$

If M is a bound for h, we let $\delta = \varepsilon / M$. Then if $\nu(E) < \delta$,

$$\int_E f d\nu < \int_E h d\nu + \varepsilon < 2\varepsilon .$$

Notice that Problem 18, Chapter 7 provides another proof of Proposition 6. ▥

Problem 11. Finish the proof of Proposition 6 for f which are not non-negative. ◀

Now we know that if f is a ν-integrable non-negative function, and we define β by

$$\beta(E) = \int_E f\,d\nu, \tag{1}$$

then β is a finite measure, and $\beta(E) \longrightarrow 0$ as $\nu(E) \longrightarrow 0$. We will say that a measure β is **absolutely continuous** with respect to a measure ν, written $\beta << \nu$, if given any $\varepsilon > 0$ there is $\delta > 0$ so that $\beta(E) < \varepsilon$ whenever $\nu(E) < \delta$. Hence if f is non-negative and integrable, and β is defined by (1) as the integral of f, then β is absolutely continuous with respect to ν. We will show that all absolutely continuous measures arise as integrals, but first we give another characterization of the absolute continuity relation $\beta << \nu$.

Proposition 7. *If β, ν are two finite measures on X, then $\beta << \nu$ if and only if $\beta(E) = 0$ whenever $\nu(E) = 0$.*

Proof. Suppose first that $\beta << \nu$. Suppose $\nu(E) = 0$. Given $\varepsilon > 0$ there is $\delta > 0$ so that $\nu(E) < \delta$ implies $\beta(E) < \varepsilon$. Since $\nu(E) < \delta$ for all δ, $\beta(E) < \varepsilon$ for all ε; *i.e.*, $\beta(E) = 0$ if $\nu(E) = 0$.

Now assume that $\nu(E) = 0$ implies $\beta(E) = 0$. Suppose β is not absolutely continuous with respect to ν, so there is $\varepsilon > 0$ and a sequence $\{E_n\}$ of measurable sets such that $\nu(E_n) < 1/2^n$ and $\beta(E_n) \geq \varepsilon$. Let

$$E = \bigcap_n (E_n \cup E_{n+1} \cup \cdots)$$

so that for all n,

$$\nu(E) \leq \sum_{i=n}^{\infty} \nu(E_i) \leq 1/2^{n-1}.$$

Hence $\nu(E) = 0$. On the other hand, if

$$F_n = E_n \cup E_{n+1} \cup \cdots,$$

then $F_1 \supset F_2 \supset \cdots$ and $\bigcap F_n = E$. Since β is a finite measure,

$$\beta(E) = \lim \beta(F_n) \geq \liminf \beta(E_n) \geq \varepsilon\,.$$

Thus there is E with $\nu(E) = 0$ and $\beta(E) \neq 0$ if β is not absolutely continuous with respect to ν. ▮

Problem 12. Let $\{\mu_n\}$ be a sequence of measures on the same space X and σ-algebra $\mathcal{S}$, and assume $\mu_n(X) \leq 1$ for all n. Let $\mu(E) = \sum \mu_n(E)/2^n$ for every measurable set E. Show that μ is a measure on X and $\mu_n << \mu$ for all n. ◀

If β and ν are signed measures, then we say $\beta << \nu$ provided $|\beta| << |\nu|$, where $\nu = \nu^+ - \nu^-$ is the Jordan decomposition of ν, and $|\nu| = \nu^+ + \nu^-$.

Problem 13. Show that if β, ν are finite signed measures, then $\beta << \nu$ if and only if $|\nu|(E) = 0$ implies $|\beta|(E) = 0$. ◀

Our goal is to show that if $\beta << \nu$ for positive finite measures β and ν, then there is a non-negative integrable function f such that

$$\beta(E) = \int_E f d\nu \tag{1}$$

for all measurable sets E. Notice that if β is the integral of f as in (1), and $\beta \not\equiv 0$, then some set $A_n = \left\{x : \frac{1}{2^n} \leq f(x) < \frac{1}{2^{n-1}}\right\}$ has positive ν-measure; so

$$\beta(A_n) = \int_{A_n} f d\nu \geq \frac{1}{2^n}\nu(A_n)\,.$$

Therefore there is a set A $(= A_n)$ and $\varepsilon > 0$ $(\varepsilon = \frac{1}{2^n})$ so that $\nu(A) > 0$ and for every subset $E \subset A$,

$$\beta(E) - \varepsilon\nu(E) \geq 0\,.$$

We show next as a necessary lemma that this condition holds whenever $\beta << \nu$.

Proposition 8. *If β and ν are finite measures and $\beta \not\equiv 0$ and $\beta << \nu$, then there is a set E_0 of positive ν-measure and $\varepsilon > 0$ so that $\beta(E) - \varepsilon\nu(E) \geq 0$ for all $E \subset E_0$.*

Proof. For every n, let A_n, B_n be a Hahn decomposition for the signed measure $\beta - \frac{1}{n}\nu$. Thus for all n,

$$\beta(B_n) \leq \frac{1}{n}\nu(B_n)\,.$$

Let $B = \bigcap B_n$, $A = \bigcup A_n$. Then for all n

$$\beta(B) \leq \beta(B_n) \leq \frac{1}{n}\nu(B_n) \leq \frac{1}{n}\nu(X)\,,$$

so $\beta(B) = 0$. Since $A \cup B = X$ and $\beta \not\equiv 0$, $\beta(A) > 0$. Since $\beta << \nu$, it follows that $\nu(A) > 0$, and consequently $\nu(A_n) > 0$ for some n. Since A_n is a positive set for $\beta - \frac{1}{n}\nu$,

$$\beta(E) - \frac{1}{n}\nu(E) \geq 0$$

for all subsets $E \subset A_n$. This is the desired result with $E_0 = A_n$ and $\varepsilon = \frac{1}{n}$. ||||||

Proposition 9. *(Radon-Nikodym Theorem). If β and ν are finite measures on X and $\beta << \nu$, then there is a non-negative ν-integrable function f on X such that*

$$\beta(E) = \int_E f\,d\nu$$

for every measurable set E.

Proof. Assume ν and β are finite measures on the same σ-algebra of subsets of X, and $\beta << \nu$. Let $\mathcal{F}$ be the set of all measurable non-negative functions f such that

$$\int_E f\,d\nu \leq \beta(E)$$

for every measurable set E. Notice that there are non-zero functions $f \in \mathcal{F}$ by Proposition 8; *e.g.*, $f = \varepsilon\chi_{E_0}$. We will show that if

$f_1, f_2 \in \mathcal{F}$, then $f_1 \vee f_2 \in \mathcal{F}$. To see this, let E be any measurable set and let

$$E_1 = \{x \in E : f_1(x) \geq f_2(x)\}, \quad E_2 = E - E_1 .$$

Then

$$\begin{aligned}\int_E f_1 \vee f_2 d\nu &= \int_{E_1} f_1 d\nu + \int_{E_2} f_2 d\nu \\ &\leq \beta(E_1) + \beta(E_2) = \beta(E) .\end{aligned}$$

Let

$$p = \sup\left\{\int_X f d\nu : f \in \mathcal{F}\right\} > 0 .$$

Since $\beta(X) < \infty$, it follows that $p < \infty$. Select functions $f_n \in \mathcal{F}$ such that

$$\int_X f_n d\nu \longrightarrow p .$$

Let $g_n = f_1 \vee f_2 \vee \cdots \vee f_n$, so $g_n \in \mathcal{F}$ and $\int g_n d\nu$ increases to p. Therefore, for fixed n and each measurable set E,

$$\int_E g_n d\nu \leq \beta(E) .$$

Let $f = \sup f_n = \lim g_n$. For any set E,

$$\int_E f d\nu = \lim_n \int_E g_n d\nu \leq \beta(E) .$$

The function f is a candidate for the largest function such that $\int_E f d\nu \leq \beta(E)$ for all E, and we want to show that equality holds. Let α be the positive measure defined by

$$\alpha(E) = \beta(E) - \int_E f d\nu .$$

Clearly α is a finite measure, and $\alpha << \nu$. By Proposition 8, either $\alpha \equiv 0$, which is what we want, or there is a set E_0 of positive ν measure and $\varepsilon > 0$ so that E_0 is a positive set for $\alpha - \varepsilon\nu$. In the latter case it follows that

$$\alpha(E_0) \geq \int \varepsilon \chi_{E_0} d\nu ,$$

or

$$\beta(E_0) - \int_{E_0} f\,d\nu \geq \int_{E_0} \varepsilon\,d\nu\,,$$
$$\beta(E_0) \geq \int_{E_0} (f + \varepsilon)\,d\nu\,.$$

Since $f + \varepsilon\chi_{E_0} \in \mathcal{F}$ and $\int_X (f + \varepsilon\chi_{E_0})d\nu > p$ we have a contradiction. ▮

The last few results above all hold equally well for σ-finite measures. The Radon-Nikodym Theorem for σ-finite measures is stated next, and the proof is left to the problems.

Proposition 10. *If β and ν are σ-finite measures on X and $\beta << \nu$, then there is a unique non-negative measurable function f on X such that $\beta(E) = \int_E f\,d\nu$ for all measurable sets E.*

The function f of Propositions 9 and 10 is called the **Radon-Nikodym derivative** of β with respect to ν, with the notation

$$d\beta = f\,d\nu, \text{ or } f = \frac{d\beta}{d\nu}\,.$$

Notice that the function f of Proposition 10 will not be integrable if $\beta(X) = \infty$.

Problem 14. Prove Proposition 10. ◀

Problem 15. Let β be Lebesgue measure on $[0,1]$, so β is defined on the σ-algebra $\mathcal{S}$ of measurable subsets of $[0,1]$. Let ν be counting measure on the same σ-algebra $\mathcal{S}$; *i.e.*, if $E \in \mathcal{S}$, $\nu(E)$ is the number of elements in E, so $\nu(E)$ is usually $+\infty$.

Obviously $\beta << \nu$. Show that there is no measurable f on $[0,1]$ such that $\beta(E) = \int_E f\,d\nu$. What hypotheses are missing from Proposition 10, and where does the proof fail? ◀

Problem 16. State and prove a Radon-Nikodym Theorem for signed measures. ◀

The Radon-Nikodym Theorem characterizes measures β such that $\beta(E) = 0$ whenever $\nu(E) = 0$. To describe the antipodal relationship between β and ν it is helpful to think of measures as distributions of mass over the set X. Thus the Lebesgue measure μ on $\mathbb{R}$ can be thought of as a uniform mass distribution where mass is the same as length. A point mass at $x_0 \in (0,1)$ would be the measure β defined on measurable sets E by $\beta(E) = 1$ if $x_0 \in E$ and $\beta(E) = 0$ if $x_0 \notin E$. Thus all the mass of β is concentrated on the single point x_0, or more exactly, on the single set $\{x_0\}$, which has Lebesgue measure zero.

Recall that two measures β and ν are **singular** with respect to each other if there exist disjoint measurable sets A and B such that $A \cup B = X$ and $\nu(A) = \beta(B) = 0$. For example, Lebesgue measure on $(0,1)$ and point mass at $x_0 = \frac{1}{2}$ are mutually singular. We will also say β is singular with respect to ν or ν is singular with respect to β depending on the emphasis. The notation is $\beta \perp \nu$.

Proposition 11. *If β and ν are two σ-finite measures on X (i.e., on the same σ-algebra of subsets of X), then β can be written as a sum $\beta = \beta_0 + \beta_1$ with $\beta_0 << \nu$ and $\beta_1 \perp \nu$.*

Proof. Let $\lambda = \beta + \nu$, so that λ is σ-finite and both β and ν are absolutely continuous with respect to λ. Let f be a non-negative measurable function on X such that

$$\nu(E) = \int_E f \, d\lambda$$

for all measurable sets E. Let

$$A = \{x : f(x) > 0\}, \quad B = \{x : f(x) = 0\},$$

so A and B are disjoint measurable sets and $X = A \cup B$. We define

$$\beta_0(E) = \beta(E \cap A); \quad \beta_1(E) = \beta(E \cap B).$$

Clearly $\beta = \beta_0 + \beta_1$. To see that $\beta_1 \perp \nu$ observe that

$$\beta_1(A) = \beta(A \cap B) = \beta(\varnothing) = 0,$$
$$\nu(B) = \int_B f \, d\lambda = \int_B 0 \, d\lambda = 0.$$

To see that $\beta_0 << \nu$, let $\nu(E) = 0$. Then $f = 0$ a.e. $[\lambda]$ on E; *i.e.*,

$$\lambda\{x \in E : f(x) \neq 0\} = 0\,,$$
$$\lambda\{x \in E : f(x) > 0\} = 0\,,$$
$$\lambda(E \cap A) = 0\,.$$

Since

$$\beta_0(E) = \beta(E \cap A) \leq \lambda(E \cap A) = 0\,,$$
$$\beta_0(E) = 0 \text{ if } \nu(E) = 0. \quad ||||||$$

Problem 17. If β and ν are finite measures on X, and $\beta << \nu$ and $\beta \perp \nu$, then $\beta(E) = 0$ for all E. ◂||||

Problem 18. Suppose β and ν are finite measures on X, and $\beta << \nu$. Let $\lambda = \beta + \nu$, so $\beta << \lambda$, and hence $\beta(E) = \int_E g\,d\lambda$ for some $g \geq 0$ and all measurable sets E. Show that $0 \leq g < 1$ almost everywhere with respect to ν; *i.e.*, $g(x) < 1$ except possibly on a set E of ν-measure zero. ◂||||

Problem 19. Let $\mathcal{S}$ and $\mathcal{T}$ be σ-algebras on the same set X, with $\mathcal{T} \subset \mathcal{S}$. Let ν be a finite measure on $\mathcal{S}$ and let f be a non-negative ν-integrable function. Let ν_0 be the restriction of ν to $\mathcal{T}$. The function f is necessarily $\mathcal{S}$-measurable but not necessarily $\mathcal{T}$-measurable, and therefore not necessarily ν_0-integrable. Show that there is a $\mathcal{T}$-measurable, ν_0-integrable function g on X such that $\int_A g\,d\nu_0 = \int_A f\,d\nu$ for all $A \in \mathcal{T}$. ◂||||

14 PRODUCT MEASURES

Let $(X, \mathcal{S}, \mu)$ and $(Y, \mathcal{J}, \nu)$ be two measure spaces. We want to define the product measure λ on $Z = X \times Y$ so that sets $A \times B$ have measure $\lambda(A \times B) = \mu(A)\nu(B)$. The process is roughly parallel to that used to develop plane measure λ in Chapters 9 and 10. Perhaps the most useful result of this chapter is the theorem which allows one to change the order of integration in an iterated integral:

$$\int_X \left(\int_Y f(x,y)\,d\nu(y) \right) d\mu(x) = \int_Y \left(\int_X f(x,y)\,d\mu(x) \right) d\nu(y)\,.$$

The most effective path to this result involves showing that both iterated integrals equal the "double" integral with respect to the product measure λ:

$$\int_Z f(x,y)\,d\lambda(x,y)\,.$$

In many cases the product integral above is little more than a curiosity except for its relationship with the iterated integrals.

Plane measure λ on $\mathbb{R}^2$ provides our model in a limited sense. Recall that the outer measure of a set $E \subset \mathbb{R}^2$ is the inf of all sums $\sum \alpha(R_i)$, where $\{R_i\}$ is a sequence of rectangles which cover E, and $\alpha(R_i)$ is the area. One of the first problems with plane measure was to show that the outer measure of a rectangle is its area; *i.e.*, that outer measure really extends the basic idea of area for rectangles. The same problem had to be faced for linear measure; namely, to show that the outer measure of an interval is its length. For both linear and plane measure we could use compactness to reduce some countable coverings to finite coverings, and thereby simplify the arguments. In the general setting of this chapter we have no such simplification, which accounts for the distressing amount of set theory which is necessary.

Our plan in outline is as follows. We will call a set $A \times B$, with A a measurable subset of X and B a measurable subset of Y, a **rectangle**. We will define the "area" of a rectangle by

$$\alpha(A \times B) = \mu(A)\nu(B)\,,$$

and the area of a finite disjoint union of rectangles as the sum of their areas; thus, if $E = A_1 \times B_1 \cup \cdots \cup A_n \times B_n$ and the $A_i \times B_i$ are disjoint, then we define

$$\alpha(E) = \sum_{i=1}^{n} \alpha(A_i \times B_i) = \sum_{i=1}^{n} \mu(A_i)\nu(B_i)\,. \tag{1}$$

The problem immediately arises to show that (1) gives a consistent definition, since the representation of E as a finite disjoint union is not unique; indeed, a rectangle can be written as a finite disjoint union of rectangles, and in many different ways.

Once we have shown that α is consistently defined by (1) on the set R of all finite disjoint unions of rectangles, we show that R is an **algebra** of sets; *i.e.*, a family of sets which is closed under finite unions and finite intersections, and differences and complementation. Since α is consistently defined on R, α is finitely additive on R. R is not necessarily a σ-algebra, so countable unions of sets in R need not belong to R. However, a countable disjoint union of sets of R might be in R, and if so, say $E = \bigcup E_i \in R$, we need to know that α is countably additive for this family:

$$\alpha\left(\bigcup E_i\right) = \sum \alpha(E_i)\,. \tag{2}$$

Equation (2) is essential if $\bigcup E_i \in R$ because our final measure λ is to be an extension of α, so of course (2) must hold for α if it is to hold for λ.

Once we have α defined on the algebra R, with α countably additive on R in the above sense, we define λ on all subsets of Z in terms of countable coverings by sets in R:

$$\lambda(E) = \inf\left\{\sum \alpha(E_i) : E \subset \bigcup E_i, E_i \in R\right\}.$$

Measurability with respect to λ is defined by the Carathéodory criterion, and as before the measurable sets form a σ-algebra. The sets in R must be shown to be measurable, and then we know that the σ-algebra of measurable sets contains the σ-algebra generated by the rectangles $A \times B$.

Now the details. Throughout this chapter $(X, \mathcal{S}, \mu)$ and $(Y, \mathcal{J}, \nu)$ will be measure spaces, and $Z = X \times Y$. All subsets $A \times B$ of Z with A and B measurable subsets of X and Y respectively are called **rectangles**, or sometimes **measurable rectangles**. We let R denote all sets in Z which are finite unions of rectangles. We define α on rectangles by $\alpha(A \times B) = \mu(A)\nu(B)$.

Proposition 1. *Every set in R can be written as a finite disjoint union of rectangles. R is an algebra; i.e., R is closed under finite unions and finite intersections, and differences and complementation.*

Proof. The two dimensional picture (of genuine rectangles) shows that the union, intersection, or difference of any two rectangles can be written as a finite disjoint union of rectangles. The complement of a rectangle is a finite disjoint union of rectangles, since

$$(A \times B)' = (A' \times B') \cup (A' \times B) \cup (A \times B').$$

To show by induction that any finite union of rectangles is a finite disjoint union, suppose this is true for any n rectangles. Let $R_1, \dots, R_{n+1}$ be $n+1$ rectangles. Let

$$R_1 \cup \cdots \cup R_n = S_1 \cup \cdots \cup S_m$$

where the S_i are disjoint rectangles. Then each set $S_i - R_{n+1}$ is a finite disjoint union of rectangles, so

$$R_1 \cup \cdots \cup R_{n+1} = R_{n+1} \cup (S_1 - R_{n+1}) \cup \cdots \cup (S_m - R_{n+1})$$

and the right side is a finite disjoint union of rectangles. Of course any finite intersection of rectangles is again a rectangle. If $E_1, E_2 \in R$,

$$E_1 = R_1 \cup \cdots \cup R_n,$$
$$E_2 = S_1 \cup \cdots \cup S_m,$$

with the R_i and the S_j disjoint rectangles, then the $R_i \cap S_j$ are disjoint rectangles and

$$E_1 \cap E_2 = \bigcup_{i,j} R_i \cap S_j\,,$$

so R is closed under finite intersections. ▮

Problem 1. Show that the intersection of any family of algebras is again an algebra. Hence there is a smallest algebra of sets containing the rectangles; this is called the algebra generated by the rectangles. ◀

Now we want to show that α is countably additive on R. The critical step is the next proposition.

Proposition 2. *If $A \times B$ is a finite or countable union of disjoint rectangles, $A \times B = \bigcup A_i \times B_i$, then*

$$\alpha(A \times B) = \sum \alpha(A_i \times B_i)\,.$$

Proof. The characteristic function of $E = \bigcup A_i \times B_i = A \times B$ is

$$\chi_E(x, y) = \chi_A(x)\chi_B(y) = \sum \chi_{A_i}(x)\chi_{B_i}(y)\,. \tag{3}$$

Integrate both sides of (3) with respect to ν, using the Monotone Convergence Theorem in the case of a countable sum:

$$\chi_A(x)\nu(B) = \sum \chi_{A_i}(x)\nu(B_i)\,. \tag{4}$$

Now integrate both sides of (4) with respect to μ, again using the Monotone Convergence Theorem for a countable sum:

$$\mu(A)\nu(B) = \sum \mu(A_i)\nu(B_i)\,.$$

This is the same as

$$\alpha(A \times B) = \sum \alpha(A_i \times B_i)\,. \quad ▮$$

Proposition 3. *If $E \in R$ and E is written as a finite or countable disjoint union of rectangles in two different ways,*

$$E = \bigcup R_i = \bigcup S_j ,$$

then $\sum \alpha(R_i) = \sum \alpha(S_j)$.

Proof. Each R_i and S_j can be written as a disjoint union of rectangles as follows:

$$R_i = \bigcup_j R_i \cap S_j ,$$
$$S_j = \bigcup_i R_i \cap S_j .$$

By Proposition 2,

$$\alpha(R_i) = \sum_j \alpha(R_i \cap S_j) ,$$
$$\alpha(S_j) = \sum_i \alpha(R_i \cap S_j) .$$

Therefore,

$$\sum_i \alpha(R_i) = \sum_j \alpha(S_j) = \sum_{i,j} \alpha(R_i \cap S_j) . \quad ▮$$

Now we can define α unambiguously on the algebra R by

$$\alpha(E) = \sum \alpha(R_i)$$

for any finite disjoint family $\{R_i\}$ of rectangles whose union is E. Moreover, using Proposition 3, it is easy to verify that α is countably additive on R.

Problem 2. Show that α is countably additive on R; *i.e.*, if $E_i = \bigcup_j R_{ij} \in R$, where $\{R_{ij}\}$ is a finite collection of disjoint rectangles, and $E = \bigcup E_i \in R$, where $\{E_i\}$ is a countable disjoint family, then $\alpha(E) = \sum \alpha(E_i)$. ◀

The outer measure λ is defined on all subsets of Z by

$$\lambda(E) = \inf\left\{\sum \alpha(E_i) : E \subset \bigcup E_i, E_i \in R\right\},$$

where the inf is over all countable coverings $\{E_i\}$. This differs formally from the definition of plane measure in Chapter 9 in that here we use coverings by sets which are finite unions of rectangles (*i.e.*, sets in R) rather than coverings by rectangles. The difference is not material because α is additive on the sets in R. The formal change to R rather than sticking to rectangles $A \times B$ is desirable because R is an algebra of sets, and we will use this fact.

Problem 3. Verify that λ is an outer measure; *i.e.*, that $\lambda(\emptyset) = 0$, and λ is a monotone, non-negative, and countably subadditive function defined on all subsets of Z. ◀

As before, a set $E \subset Z$ is **measurable** provided

$$\lambda(E \cap T) + \lambda(E' \cap T) = \lambda(T)$$

for every set T. The proofs of Chapter 4 apply without change to show that the λ-measurable sets form a σ-algebra, and that λ is countably additive on this σ-algebra.

Proposition 4. *The sets in R are measurable; in particular, rectangles are measurable.*

Proof. (cf. Proposition 4, Chapter 9) Let $E \in R$. We must show that

$$\lambda(E \cap T) + \lambda(E' \cap T) \le \lambda(T)$$

for every set T. Assume that $\lambda(T) < \infty$ and let $\{E_i\}$ be a countable covering of T by sets from R, with

$$\sum \alpha(E_i) < \lambda(T) + \varepsilon .$$

Then

$$E \cap T \subset \bigcup (E_i \cap E),$$
$$E' \cap T \subset \bigcup (E_i \cap E'),$$

and since R is an algebra, the sets $E \cap E_i$, $E' \cap E_i$ are all in R. Therefore, since α is additive on R,

$$\begin{aligned}\lambda(E \cap T) + \lambda(E' \cap T) &\le \sum \alpha(E \cap E_i) + \sum \alpha(E' \cap E_i) \\ &= \sum [\alpha(E \cap E_i) + \alpha(E' \cap E_i)] \\ &= \sum \alpha(E_i) \\ &< \lambda(T) + \varepsilon\,.\end{aligned}$$

Since ε is arbitrary, E is measurable. ||||||

We will let R_σ denote all sets which can be written as countable unions of sets in R, and $R_{\sigma\delta}$ will denote all countable intersections of sets in R_σ. We will find for every measurable set E a set A in R_σ which includes E and has approximately the same measure. We then find a set B in $R_{\sigma\delta}$ which includes E and has the same measure as E. We can show that sets in R_σ, and then $R_{\sigma\delta}$, have the basic property relating λ to μ and ν:

$$\lambda(A) = \int \nu(A_x)\,d\mu(x)\,.$$

We then get the result for arbitrary measurable sets E by limit theorems on the integrals, and from this we can show the iterated integrals are equal. First we need the following lemma.

Proposition 5. *Every set $A \in R_\sigma$ can be written as a union of increasing sets in R, and every set $B \in R_{\sigma\delta}$ can be written as the intersection of a decreasing sequence of sets in R_σ.*

Proof. Let $A \in R_\sigma$, with $A = \bigcup E_i$ and each $E_i \in R$. If $F_n = E_1 \cup \cdots \cup E_n$, then $F_n \in R$ and $F_1 \subset F_2 \subset \cdots$, and $E = \bigcup F_n$.

Now let A_1 and A_2 be sets of R_σ, with

$$A_1 = \bigcup E_{1i}\,, \quad A_2 = \bigcup E_{2j}\,,$$

and all the sets $E_{1i}, E_{2j} \in R$. Then

$$A_1 \cap A_2 = \bigcup_{i,j} (E_{1i} \cap E_{2j})\,,$$

and all the sets $E_{1i} \cap E_{2j}$ are in R, so $A_1 \cap A_2 \in R_\sigma$. Hence R_σ is closed under finite intersections. If $B \in R_{\sigma\delta}$, with $B = \bigcap A_n$ and each $A_n \in R_\sigma$, then let

$$A_n^* = A_1 \cap \cdots \cap A_n ,$$

so each $A_n^* \in R_\sigma$, $A_1^* \supset A_2^* \supset \cdots$, and $B = \bigcap A_n^*$. ▮

Proposition 6. *If E is any measurable set with $\lambda(E) < \infty$, and $\varepsilon > 0$, there is a set $A \in R_\sigma$ with $E \subset A$ and $\lambda(A) < \lambda(E) + \varepsilon$. There is also a set $B \in R_{\sigma\delta}$ with $E \subset B$ and $\lambda(B) = \lambda(E)$.*

Proof. Let $\{E_i\}$ be a countable covering of E by sets in R with

$$\sum \lambda(E_i) < \lambda(E) + \varepsilon .$$

If $A = \bigcup E_i$, then $A \in R_\sigma$, $E \subset A$, and

$$\lambda(A) \le \sum \lambda(E_i) < \lambda(E) + \varepsilon .$$

Now let $A_n \in R_\sigma$ with $E \subset A_n$ and $\lambda(A_n) < \lambda(E) + \frac{1}{n}$. We may assume that $A_1 \supset A_2 \supset \cdots$. Let $B = \bigcap A_n$, so $B \in R_{\sigma\delta}$ and $E \subset B$. Since $\lambda(A_1) < \infty$ and the A_n are nested, $\lambda(B) = \lim \lambda(A_n) = \lambda(E)$. ▮

Proposition 7. *All sections of sets in $R_{\sigma\delta}$ are measurable; i.e., if $B \in R_{\sigma\delta}$ then $B_x = \{y : (x, y) \in B\}$ is a measurable subset of Y for each x and $B^y = \{x : (x, y) \in B\}$ is a measurable subset of X for each y. Since $R \subset R_\sigma \subset R_{\sigma\delta}$, the sections of R and R_σ sets are measurable.*

Proof. If E is a rectangle, or a finite union of rectangles (*i.e.*, a set in R) then sections are obviously measurable. If $A \in R_\sigma$, with $A = \bigcup E_i$, $E_i \in R$, then for each x

$$A_x = \left(\bigcup E_i\right)_x = \bigcup (E_i)_x ,$$

so A_x is measurable. If $B \in R_{\sigma\delta}$ with $B = \bigcap A_n$ and each $A_n \in R_\sigma$, then each $(A_n)_x$ is measurable so

$$B_x = \left(\bigcap A_n\right)_x = \bigcap (A_n)_x$$

is measurable. ▮

Proposition 8. *If $B \in R_{\sigma\delta}$ and $\lambda(B) < \infty$, and*

$$f(x) = \nu(B_x), \quad g(y) = \mu(B^y),$$

then f and g are bounded measurable functions, and

$$\int f(x)d\mu(x) = \int g(y)d\nu(y) = \lambda(B).$$

Proof. If B is a rectangle, the result is clear. If $B \in R_\sigma$, then B is a countable disjoint union of rectangles, $B = \bigcup R_n$. Hence

$$f(x) = \nu(E_x) = \sum \nu(R_n)_x,$$

so f is the sup of a sequence of measurable functions. Moreover, by the Monotone Convergence Theorem,

$$\begin{aligned}\int f(x)d\mu(x) &= \sum \int \nu(R_n)_x d\mu(x) \\ &= \lim_n \sum_{i=1}^{n} \lambda(R_i) \\ &= \lim_n \lambda\left(\bigcup_{i=1}^{n} R_i\right) \\ &= \lambda\left(\bigcup_{i=1}^{\infty} R_i\right) = \lambda(B).\end{aligned}$$

Now let $B \in R_{\sigma\delta}$, with $\lambda(B) < \infty$. Then $B = \bigcap A_n$ with $A_n \in R_\sigma$ and we can take $A_1 \supset A_2 \supset \cdots$ and $\lambda(A_1) < \infty$. We know

$$\int \nu(A_1)_x d\mu(x) = \lambda(A_1) < \infty,$$

so $\nu(A_1)_x$ is measurable and finite valued for almost all x. For these x, $\nu(A_n)_x$ is a decreasing sequence of integrable functions and, since $\nu(A_1)_x$ is integrable,

$$\int \nu(A_n)_x d\mu(x) \longrightarrow \int \nu(B_x)d\mu(x).$$

But

$$\int \nu(A_n)_x d\mu(x) = \lambda(A_n)$$

and

$$\lambda(A_n) \longrightarrow \lambda(B),$$

so

$$\lambda(B) = \int \nu(B_x) d\mu(x). \quad ▮$$

To show that

$$\lambda(E) = \int \nu(E_x) d\mu(x)$$

holds for all measurable sets E of finite measure, we surround E with a $R_{\sigma\delta}$ set B with the same measure. The result follows once we show that $\nu(B_x) = \nu(E_x)$ a.e.

Proposition 9. *If E is measurable and $\lambda(E) < \infty$, then E_x and E^y are measurable for almost all x and almost all y, and*

$$f(x) = \nu(E_x), \quad g(y) = \mu(E^y)$$

are measurable functions defined a.e., and

$$\lambda(E) = \int f(x) d\mu(x) = \int g(y) d\nu(y).$$

Proof. Let $\lambda(E) < \infty$ and select $B \in R_{\sigma\delta}$ with $E \subset B$ and $\lambda(B) = \lambda(E)$. Let $C = B - E$, so C is measurable and $\lambda(C) = 0$. For all x, $C_x = B_x - E_x$. Let $H \in R_{\sigma\delta}$ where $H \supset C$ and $\lambda(H) = \lambda(C) = 0$. Then

$$\int \nu(H_x) d\mu(x) = \lambda(H) = 0.$$

It follows that $\nu(H_x) = 0$ a.e., and since $H_x \supset C_x$, $\nu(C_x) = 0$ a.e. Therefore $E_x = B_x$ a.e., so $\nu(E_x)$ is measurable and

$$\lambda(E) = \int \nu(E_x) d\mu(x).$$

The companion result for y-sections is proved similarly. ▮

Proposition 10. *(Fubini) Let $(X, \mathcal{S}, \mu)$ and $(Y, \mathcal{J}, \nu)$ be two measure spaces, and let λ be the product measure on $X \times Y$. For any integrable function $f(x, y)$ on $X \times Y$ let*

$$f_x(y) = f(x, y), \quad f^y(x) = f(x, y).$$

Then f_x is an integrable function on Y for almost all x and f^y is an integrable function on X for almost all y, and

$$F(x) = \int_Y f(x, y) d\nu(y), \quad G(y) = \int_X f(x, y) d\mu(x)$$

are integrable functions on X and Y respectively, and

$$\int_X F(x) d\mu(x) = \int_Y G(y) d\nu(y) = \int_{X \times Y} f(x, y) d\lambda(x, y);$$

i.e.,

$$\int_X \left(\int_Y f(x, y) d\nu(y) \right) d\mu(x) = \int_Y \left(\int_X f(x, y) d\mu(x) \right) d\nu(y)$$
$$= \int_{X \times Y} f(x, y) d\lambda(x, y).$$

Proof. By symmetry we need only show that f_x is integrable for almost all x, and $F(x) = \int f_x(y) d\nu(y)$ is integrable, and

$$\int_X F(x) d\mu(x) = \int_{X \times Y} f(x, y) d\lambda(x, y).$$

It is sufficient (Problem 4 below) to consider only non-negative functions $f(x, y)$.

Since $f(x, y)$ is integrable and non-negative, there is an increasing sequence $\{\varphi_n\}$ of simple functions on $X \times Y$ so that $\varphi_n(x, y) \longrightarrow f(x, y)$ a.e. $[\lambda]$. It follows that

$$(\varphi_n)_x(y) \longrightarrow f_x(y) \qquad \text{a.e.}[\lambda].$$

If φ is simple, φ is a finite sum of functions $\psi(x, y)$ of the form

$$\psi(x, y) = a \chi_E(x, y)$$

where E is measurable and $\lambda(E) < \infty$. Notice that

$$\psi_x(y) = a\chi_{E_x}(y).$$

Since E_x is a measurable subset of Y for almost all x, χ_{E_x} is a measurable function of y for almost all x, and each summand $\psi_x(y)$ of $\varphi_x(y)$ is measurable. Therefore φ_x is measurable for almost all x if φ is a simple function. If $\varphi_n(x,y)$ increases to $f(x,y)$, then $(\varphi_n)_x$ increases to f_x, so f_x, is measurable for almost all x. By the Monotone Convergence Theorem,

$$\int (\varphi_n)_x(y)d\nu(y) \longrightarrow \int f_x(y)d\nu(y) = F(x)$$

for almost all x. Since each φ_n is simple, each integral on the left is a measurable function of x by Proposition 9, so F is a measurable function. By the Monotone Convergence Theorem (twice), and Proposition 9 again,

$$\begin{aligned}\int_X \left(\int_Y f(x,y)d\nu(y)\right) d\mu(x) &= \lim \int_X \left(\int_Y \varphi_n d\nu\right) d\mu \\ &= \lim \int_{X\times Y} \varphi_n(x,y)d\lambda(x,y) \\ &= \int_{X\times Y} f(x,y)d\lambda(x,y). \quad ▮\end{aligned}$$

Problem 4. Show that it suffices to prove the Fubini Theorem for non-negative functions. ▮

Problem 5. Let X and Y be uncountable sets, for instance $X = Y = [0,1]$. Let $\mu\{x\} = 1$ for all $x \in X$ and $\nu\{y\} = 1$ for all $y \in Y$, so both μ and ν are counting measure. If $\lambda = \mu \times \nu$, then $\lambda\{(x,y)\} = 1$ for all (x,y). Show that no sequence of simple functions increases to the function f which is identically one on $X \times Y$, so the proof of Fubini's Theorem fails. (Of course this f is not integrable.) ▮

Proposition 11. *(Tonelli's Theorem) If $(X, \mathcal{S}, \mu)$ and $(Y, \mathcal{J}, \nu)$ are σ-finite measure spaces and f is a non-negative measurable* *function on $X \times Y$, then the necessary measurability conditions hold as in the Fubini Theorem, and*

$$\int_X \left(\int_Y f(x,y) d\nu(y) \right) d\mu(x) = \int_Y \left(\int_X f(x,y) d\mu(x) \right) d\nu(y)$$
$$= \int_{X \times Y} f(x,y) d\lambda(x,y).$$

Proof. If $X = \bigcup X_n$, $Y = \bigcup Y_n$ with each X_n and Y_n of finite measure and $X_1 \subset X_2 \subset \cdots$, $Y_1 \subset Y_2 \subset \cdots$, then $X \times Y = \bigcup X_n \times Y_n$ and $X_n \times Y_n$ has finite λ-measure. There is a sequence $\{\varphi_{nj}\}$ of simple functions supported on $X_n \times Y_n$ which increases to f on $X_n \times Y_n$:

$$\lim_{i \to \infty} \varphi_{ni}(x,y) = f(x,y) \text{ on } X_n \times Y_n .$$

Therefore there is a sequence $\psi_n(x,y)$ of simple functions which increases to $f(x,y)$ on $X \times Y$. The point is that the ψ_n are zero off finite measure sets, which was the critical point where Proposition 9 is used in the proof of Fubini's Theorem. The rest of the proof of Proposition 10 proceeds as before. ▮

To use Tonelli's Theorem to change the order of integration in an iterated integral

$$\int_X \int_Y f(x,y) d\mu(x) d\nu(y),$$

you first check that μ and ν are σ-finite measures. If you are an analyst, X and Y are probably both the real line or the plane, so this is no problem. (If you are a probabilist, then you need more help than you can get here; I suggest prayer.) Given that X and Y (or μ and ν) are σ-finite, you check that f is measurable. Generally this is because f is continuous. If all you have is facts like $f(x,y)$ is continuous in y for each x and measurable in x, then you are in the realm of Polish topology, which is beyond the scope of a primer. If you know that f is measurable, and either iterated integral of $|f(x,y)|$ is finite, then you can change the order.

Problem 6. If $f(x)$ is integrable on X and $g(y)$ is integrable on Y and $h(x,y) = f(x)g(y)$, then h is integrable on $X \times Y$ and

$$\int_{X\times Y} h(x,y)\,d\lambda(x,y) = \int_X f(x)\,d\mu(x) \int_Y g(y)\,d\nu(y).$$

Do you need to assume that h is non-negative or that X and Y are σ-finite? ◀

15 THE SPACE L^2

The study of Fourier series in the early 1800's gave rise to many fundamental advances in analysis. The question was this – what functions can be represented by Fourier series? It soon became clear that this question could not be answered without a better understanding of the basic ideas of analysis, including what is meant by "function" and "represent." After the Lebesgue integral was introduced in 1904 the space L^2 of square integrable functions and the space ℓ^2 of square summable sequences emerged as heroes.

We will start with ℓ^2, which we introduce as the infinite dimensional analogue of Euclidean space.

We are familiar from elementary physics and calculus with the treatment of $\mathbb{R}^2$ and $\mathbb{R}^3$ as vector spaces. If $x = (x_1, x_2)$ and $y = (y_1, y_2)$ are elements of $\mathbb{R}^2$, and a is a real number, then we define vector sum and scalar multiplication as follows:

$$x + y = (x_1 + y_1, x_2 + y_2), \tag{1}$$

$$ax = (ax_1, ax_2). \tag{2}$$

The sum, $+$, of (1) makes $\mathbb{R}^2$ an abelian group, and the scalar multiplication (2) then makes $\mathbb{R}^2$ a **vector space,** or **linear space**. Similarly, $\mathbb{R}^3$ is a linear space under coordinate-wise addition and scalar multiplication. If $x = (x_1, x_2, x_3)$ and $y = (y_1, y_2, y_3)$, then

$$\begin{aligned} x + y &= (x_1 + y_1, x_2 + y_2, x_3 + y_3), \\ ax &= (ax_1, ax_2, ax_3). \end{aligned}$$

The absolute value of a vector (point) of $\mathbb{R}$ or $\mathbb{R}^2$ or $\mathbb{R}^3$ is its distance from the origin. For $x \in \mathbb{R}$, $|x|$ is the distance from the origin, and $|x - y|$ is the distance between x and y. For x in $\mathbb{R}^2$, $|x| = \sqrt{x_1^2 + x_2^2}$ is the distance from x to the origin, and

$$|x - y| = \sqrt{(x_1 - y_1)^2 + (x_2 - y_2)^2}$$

is the distance from x to y. Each $\mathbb{R}^n$ becomes a linear space with coordinate-wise addition and scalar multiplication. For $x \in \mathbb{R}^n$ with $n \geq 3$ the distance from the origin is called the **norm** instead of the absolute value, and the notation acquires two additional vertical bars:

$$\|x\| = \sqrt{x_1^2 + \cdots + x_n^2},$$

$$\|x - y\| = \sqrt{(x_1 - y_1)^2 + \cdots + (x_n - y_n)^2}.$$

For any $\mathbb{R}^n$ the norm satisfies the following relation with scalar multiplication:

$$\|ax\| = |a|\,\|x\|.$$

The dot product or **scalar product** of vectors in $\mathbb{R}^2$ and $\mathbb{R}^3$ is familiar, and we make the analogous definition for vectors in $\mathbb{R}^n$: if $x = (x_1, \ldots, x_n)$ and $y = (y_1, \ldots, y_2)$, then

$$(x, y) = x_1 y_1 + \cdots + x_n y_n.$$

Problem 1. Verify that the following identities hold in $\mathbb{R}^n$.
(i) $a(x + y) = ax + ay$
(ii) $\|ax\| = |a|\,\|x\|$
(iii) $(x, y) = (y, x)$
(iv) $(x + y, z) = (x, z) + (y, z)$
(v) $(ax, y) = a(x, y)$
(Parts (iii), (iv), (v) say that the inner product is a linear function of each variable.) ◀

In $\mathbb{R}$, $\mathbb{R}^2$, and $\mathbb{R}^3$ the angle θ between two vectors x and y is determined by the formula

$$(x, y) = \|x\|\,\|y\| \cos\theta. \tag{3}$$

We can define the angle θ between vectors in any $\mathbb{R}^n$ by (3) provided $|(x, y)| \leq \|x\|\,\|y\|$ for all x and y. The definition of θ is not important, but the inequality is.

Problem 2. (Cauchy's inequality.) If $x = (x_1, \ldots, x_n)$ and $y = (y_1, \ldots, y_n)$, then

$$|(x, y)| \le \|x\| \, \|y\|;$$

i.e.,

$$\left| \sum_{i=1}^{n} x_i y_i \right| \le \left(\sum_{i=1}^{n} x_i^2 \right)^{\frac{1}{2}} \left(\sum_{i=1}^{n} y_i^2 \right)^{\frac{1}{2}}.$$

Hint: First show that it is sufficient to prove this when $\|x\| = \|y\| = 1$. Notice that $|ab| \le (a^2 + b^2)/2$ for all a, b, since $(a - b)^2 \ge 0$, and therefore $|x_i y_i| \le (x_i^2 + y_i^2)/2$ for each i. Summation, using $\|x\| = \|y\| = 1$, gives the result. ▮

The other elementary but important inequality for norms in $\mathbb{R}^n$ is **Minkowski's inequality**, which gives us the triangle inequality for the metric $\|x - y\|$.

Proposition 1. *If* $x = (x_1, \ldots, x_n)$ *and* $y = (y_1, \ldots, y_n)$, *then*

$$\|x + y\| \le \|x\| + \|y\|;$$

i.e.,

$$\left(\sum (x_i + y_i)^2\right)^{\frac{1}{2}} \le \left(\sum x_i^2\right)^{\frac{1}{2}} + \left(\sum y_i^2\right)^{\frac{1}{2}}.$$

Proof. $\sum (x_i + y_i)^2 = \sum x_i^2 + \sum y_i^2 + 2 \sum x_i y_i$. By Cauchy's inequality,

$$2 \sum x_i y_i \le 2 \left(\sum x_i^2\right)^{\frac{1}{2}} \left(\sum y_i^2\right)^{\frac{1}{2}},$$

so

$$\begin{aligned}\sum (x_i + y_i)^2 &\le \sum x_i^2 + 2 \left(\sum x_i^2\right)^{\frac{1}{2}} \left(\sum y_i^2\right)^{\frac{1}{2}} + \sum y_i^2 \\ &= \left(\left(\sum x_i^2\right)^{\frac{1}{2}} + \left(\sum y_i^2\right)^{\frac{1}{2}}\right)^2.\end{aligned}$$

Taking the square root of both sides gives us the inequality we wanted to prove. ▮

Problem 3. Let $d(x, y) = \|x - y\|$ be the distance between x and y in $\mathbb{R}^n$. Verify the three conditions which characterize a metric:

(i) $d(x,y) \geq 0$ and $d(x,y) = 0$ if and only if $x = y$;
(ii) $d(x,y) = d(y,x)$ for all x, y;
(iii) $d(x,y) + d(y,z) \geq d(x,z)$ for all x, y, z.

The family of all sequences $x = (x_1, x_2, \ldots, x_n, \ldots)$ again forms a linear space under coordinate-wise addition and scalar multiplication. To get a norm analogous to that for $\mathbb{R}^n$ we restrict our attention to sequences x such that $\sum x_i^2 < \infty$. This space is called ℓ^2, and in ℓ^2 we define the norm

$$\|x\| = \left(\sum x_i^2\right)^{\frac{1}{2}},$$

and the inner product

$$(x,y) = \sum x_i y_i . \tag{4}$$

Problem 4. (i) The sum in (4) converges and $|(x,y)| \leq \|x\|\,\|y\|$ for all x, $y \in \ell^2$.

(ii) Given that x, $y \in \ell^2$, show that $x + y \in \ell^2$ and that $\|x + y\| \leq \|x\| + \|y\|$.

Now fix a measure space $(X, \mathcal{S}, \nu)$, and let L^2 be the space of all square integrable functions on X; *i.e.*, $f \in L^2$ if and only if $\int f^2 d\nu < \infty$. We will primarily be interested in just two measure spaces: (i) $X = [-\pi, \pi]$, which is the natural home for the trigonometric functions $\cos kx$, $\sin kx$; and (ii) $X = \mathbb{N}$, with ν equal to counting measure. Notice that $L^2[\mathbb{N}]$ is exactly the space ℓ^2 of square summable sequences.

For our first few results we let $(X, \mathcal{S}, \nu)$ be any measure space, and define the norm and inner product in L^2 as follows:

$$\|f\| = \left(\int f^2 d\nu\right)^{\frac{1}{2}},$$

$$(f,g) = \int fg\, d\nu .$$

Proposition 2. *If f, $g \in L^2$, then (i) $f + g \in L^2$, and $\|f + g\| \leq \|f\| + \|g\|$;*

(ii) fg is integrable, so (f,g) makes sense, and $\int |fg| \leq \|f\|\,\|g\|$. Hence $|(f,g)| \leq \|f\|\,\|g\|$ for all f, g.

Proof. The function x^2 is convex, which means that

$$(\lambda x_1 + (1-\lambda)x_2)^2 \leq \lambda x_1^2 + (1-\lambda)x_2^2 \tag{5}$$

for all numbers x_1 and x_2 all $\lambda \in (0,1)$. Assume f, $g \in L^2$ and let $f_0 = f/\|f\|$, $g_0 = g/\|g\|$, so $\|f_0\| = \|g_0\| = 1$. Estimate as follows:

$$\begin{aligned}
|f(x)+g(x)|^2 &\leq (|f(x)| + |g(x)|)^2 \\
&= (\|f\|f_0(x) + \|g\|g_0(x))^2 \\
&= (\|f\| + \|g\|)^2 \left(\frac{\|f\|}{\|f\|+\|g\|} f_0(x) + \frac{\|g\|}{\|f\|+\|g\|} g_0(x) \right)^2 .
\end{aligned}$$

The second parenthesis in the last line above is a convex combination of $f_0(x)$ and $g_0(x)$, of the form (5), so

$$\begin{aligned}
&|f(x)+g(x)|^2 \\
&\quad \leq (\|f\| + \|g\|)^2 \left(\frac{\|f\|}{\|f\|+\|g\|} f_0(x)^2 + \frac{\|g\|}{\|f\|+\|g\|} g_0(x)^2 \right) \\
&\quad = (\|f\| + \|g\|)(\|f\|f_0(x)^2 + \|g\|g_0(x)^2) .
\end{aligned}$$

The right side above is integrable, so $f+g \in L^2$. Integrating both sides above and using $\|f_0\|^2 = \|g_0\|^2 = 1$, we get

$$\|f+g\|^2 \leq (\|f\| + \|g\|)^2 .$$

Part (ii) is proved just as before. Using

$$|f(x)g(x)| \leq \left(f(x)^2 + g(x)^2\right)/2 ,$$

we see that fg is integrable, and

$$\int |fg| \leq \frac{1}{2}(\|f\|^2 + \|g\|^2) . \tag{6}$$

Replace f by $f/\|f\|$ and g by $g/\|g\|$ in (6) and get

$$\frac{1}{\|f\|\,\|g\|}\int |fg| \le \frac{1}{2}\left(\left\|\frac{f}{\|f\|}\right\|^2 + \left\|\frac{g}{\|g\|}\right\|^2\right) = 1. \quad ▮$$

The nice connection between the functions in $L^2[-\pi,\pi]$ and the Fourier series with coefficients in $\ell^2 = L^2[\mathbb{N}]$ depends on the fact that every L^2 space is complete in its norm. The following apparent digression is necessary for the completeness proof. In any normed space the convergence of infinite series is defined just as for series of reals. If $\{x_n\}$ is a sequence of vectors (points in a normed space) then $\sum x_n = x$ means $\left\|\sum_{n=1}^{N} x_n - x\right\| \longrightarrow 0$ as $N \longrightarrow \infty$. The series $\sum x_n$ is **absolutely convergent** provided the real series $\sum \|x_n\|$ converges.

Proposition 3. *A normed space is complete if and only if every absolutely convergent series converges to a point of the space.*

Proof. First assume that $\{x_n\}$ is a sequence in a complete normed space and $\sum \|x_n\|$ converges. If $s_n = x_1 + \cdots + x_n$, then

$$\begin{aligned}\|s_n - s_m\| &= \|x_{m+1} + x_{m+2} + \cdots + x_n\| \\ &\le \|x_{m+1}\| + \cdots + \|x_n\|.\end{aligned}$$

Since $\sum \|x_n\|$ converges, the right side above is less than any given ε for all sufficiently large m and n. Thus $\{s_n\}$ is a Cauchy sequence in a complete space, so s_n converges; that is, $\sum x_n$ converges.

Now assume that every absolutely convergent series converges. Let $\{x_n\}$ be a Cauchy sequence; we must show that $\{x_n\}$ converges. For each k pick n_k so that $\|x_i - x_j\| < 2^{-k}$ if $i, j \ge n_k$. We can assume that the n_k are strictly increasing. Consider the series

$$x_{n_1} + (x_{n_2} - x_{n_1}) + (x_{n_3} - x_{n_2}) + \cdots, \tag{7}$$

whose kth partial sum is x_{n_k}. This series is absolutely summable since $\|x_{n_{k+1}} - x_{n_k}\| < 1/2^k$. Therefore the series (7) converges; *i.e.*, there is some x such that $x_{n_k} \longrightarrow x$ as $k \longrightarrow \infty$. Since $\{x_n\}$ is Cauchy and has a subsequence which converges to x, the sequence $\{x_n\}$ converges to x, and the space is complete. ▮

Proposition 4. *(The Riesz-Fischer Theorem.) The space L^2 is complete. Since ℓ^2 is L^2 for $X = \mathbb{N}$ and ν equal to counting measure, ℓ^2 is complete.*

Proof. We show that every absolutely summable series in L^2 is convergent. Let $\{f_n\}$ be a sequence in L^2 such that $\sum \|f_n\| = M < \infty$. Let $g_n(x) = \sum_{i=1}^{n} |f_i(x)|$, so g_n is the sum of a finite number of L^2 functions, and hence $g_n \in L^2$. Moreover,

$$\|g_n\| \leq \sum_{i=1}^{n} \|f_i\| \leq M .$$

For each fixed x, $\{g_n(x)\}$ is an increasing sequence of real numbers, so $\{g_n\}$ converges pointwise to an extended real valued function g, which is measurable if it is finite almost everywhere. For all n,

$$\|g_n\|^2 = \int g_n^2 \leq M^2 ,$$

so $\int g^2 \leq M^2$ and $g \in L^2$. Since g^2 is integrable, g is finite almost everywhere. For those x for which $g(x) = \sum |f_i(x)|$ is finite, the series $\sum f(x)$ converges; let

$$s_n(x) = \sum_{i=1}^{n} f_i(x) ,$$

$$s(x) = \sum_{i=1}^{\infty} f_i(x) .$$

The function s is measurable. Moreover, for all n,

$$|s_n(x)| \leq \sum_{i=1}^{n} |f_i(x)| = g_n(x) ,$$

so $|s(x)| \leq g(x)$. Therefore $s \in L^2$, and the absolutely summable series $\sum f_i$ converges *pointwise* to $s \in L^2$. We must show that the series converges to s in the L^2 norm; *i.e.*, $\|s_n - s\| \longrightarrow 0$. Notice that

$$|s_n(x) - s(x)|^2 \leq (2g(x))^2 = 4g(x)^2 .$$

The functions $\left(s_n(x) - s(x)\right)^2$ converge pointwise to zero, and they are dominated by the integrable function $4g(x)^2$. Therefore

$$\int (s_n - s)^2 \longrightarrow 0;$$

i.e., $\|s_n - s\|^2 \longrightarrow 0$ and the absolutely summable series $\sum f_i$ converges to s in the L^2 norm. ▮

Now we return to Fourier series, and give one pleasant answer to the question of what functions can be represnted by Fourier series. A **trigonometric series** is a series of the form

$$\frac{1}{2}a_0 + \sum_{k=1}^{\infty} a_k \cos kx + b_k \sin kx\,. \tag{8}$$

If $s_n(x)$ is the nth partial sum of (8), then of course $s_n \in L^2[-\pi, \pi]$, and we will need the formula for the L^2 norm calculated in the following problem.

Problem 5. If $s_n(x) = \frac{1}{2}a_0 + \sum_{k=1}^{n} a_k \cos kx + b_k \sin kx$, then $\|s_n\|^2 = \pi \left[\frac{1}{2}a_0^2 + \sum_{k=1}^{n} a_k^2 + b_k^2\right]$; *i.e.*, the L^2 norm of s_n is essentially the same as the ℓ^2 norm of the sequence of coefficients: $\frac{1}{2}a_0, a_1, b_1, a_2, b_2, \ldots$. Hint: The requisite orthogonality relations are:

$$\int_{-\pi}^{\pi} \cos kx \cos \ell x \, d\mu(x) = \begin{cases} \pi \text{ if } k = \ell \\ 0 \text{ if } k \neq \ell \end{cases}$$

$$\int_{-\pi}^{\pi} \sin kx \sin \ell x \, d\mu(x) = \begin{cases} \pi \text{ if } k = \ell \\ 0 \text{ if } k \neq \ell \end{cases}$$

$$\int_{-\pi}^{\pi} \sin kx \cos \ell x \, d\mu(x) = 0. \tag{9}$$

If the trigonometric series (8) converges pointwise to an integrable function f, and if the partial sums are dominated by an integrable function so that the Lebesgue convergence theorem applies, then the coefficients $\{a_n\}$, $\{b_n\}$ are determined by f as indicated in the next problem.

Problem 6. Let $s_n(x)$ be the nth partial sum of the trigonometric series (8). If there is an integrable function g on $[-\pi,\pi]$ such that $|s_n(x)| \le g(x)$ for all x and n, and if $\{s_n\}$ converges pointwise a.e. on $[-\pi,\pi]$ to the integrable function f, then

$$a_k = \frac{1}{\pi}\int_{-\pi}^{\pi} f(x)\cos kx \, d\mu(x)\,,$$

$$b_k = \frac{1}{\pi}\int_{-\pi}^{\pi} f(x)\sin kx \, d\mu(x)\,. \quad (10)$$

Unfortunately, the kind of bounded or dominated convergence required in Problem 6 is not easy to come by, so pointwise convergence of Fourier series is less than ideal. However, the formulas (10) make sense for any integrable function f on $[-\pi,\pi]$, and hence for any L^2 function. If f is an integrable function which happens to have a trigonometric series that converges to it nicely, then we know what the series must be. Accordingly, the numbers $\{a_k\}$, $\{b_k\}$ are called the **Fourier coefficients** of f. The trigonometric series with these coefficients is the **Fourier series** for f, and no assumption is made about the convergence of the series. The mapping from integrable functions (or L^2 functions) to sequences of Fourier coefficients is linear; this is a simple consequence of the fact that the integral formulas (9) are linear functions of f. The mapping from integrable functions, and therefore from L^2-functions, to sequences of Fourier coefficients is also one-to-one; *i.e.*, if $f \neq g$, then f and g have different sequences of Fourier coefficients. The proof of this last fact is unfortunately out of our path, so this will have to be an article of faith. Given this fact, we will show that the mapping from L^2-functions to their sequences of Fourier coefficients is a one-to-one mapping on L^2 onto ℓ^2. Moreover, the mapping nearly preserves the norm in the sense that $\|f\|^2 = \pi[\frac{1}{2}a_0^2 + \sum(a_k^2 + b_k^2)]$ if $\{a_k\}$, $\{b_k\}$ are the Fourier coefficients of f. The calculations of the following problem give us most of the information we need.

Problem 7. Let $f \in L^2[-\pi,\pi]$ and let $\{a_k\}$, $\{b_k\}$ be the

Fourier coefficients of f. Let

$$P_n(x) = \frac{1}{2}\alpha_0 + \sum_{k=1}^{n} \alpha_k \cos kx + \beta_k \sin kx$$

be any nth degree trigonometric polynomial. Show that

$$\|f - P_n\|^2 = \|f\|^2 - \pi\left[\frac{1}{2}a_0^2 + \sum_{k=1}^{n} a_k^2 + b_k^2\right] + \sum_{k=1}^{n} (\alpha_k - a_k)^2 + (\beta_k - b_k)^2 . \tag{11}$$

Hint: This is nothing more than a moderately unpleasant exercise in algebra. The simplest way may be to verify the case $n = 1$, which involves completing a square, and then use induction.

Now we milk the identity (11). One obvious but significant consequence of (11) is that the nth partial sum of the Fourier series for f is *the* trigonometric polyomial of degree n or less which best approximates f in the L^2 norm. Any other nth degree trigonometric polynomial $P_n(x)$ gives a strictly larger value for $\|f - P_n\|$. We also see from (11) that if P_n is the nth partial sum of the Fourier series for f (*i.e.*, $\alpha_k = a_k$, $\beta_k = b_k$ for $k = 0, 1, \ldots, n$) then, since $\|f - P_n\|^2 \geq 0$,

$$\pi\left[\frac{1}{2}a_0^2 + \sum_{k=1}^{n} a_k^2 + b_k^2\right] \leq \|f\|^2 . \tag{12}$$

This is called **Bessel's inequality**. If follows immediately that if $f \in L^2$, the sequence $\frac{1}{2}a_0$, a_1, b_1, a_2, b_2, ... of its Fourier coefficients is in ℓ^2, and the ℓ^2-norm of this sequence is less than or equal to $\frac{1}{\sqrt{\pi}}\|f\|$. Among other things, this says that the map from functions in L^2 to their Fourier coefficients in ℓ^2 is continuous. Our final result shows that Bessel's inequality (12) is an equality, and that the mapping from L^2 to ℓ^2 is onto. That is *every* ℓ^2 sequence makes a Fourier series which represents an L^2-function, and the norms are preserved from L^2 to ℓ^2 with a constant factor, so

$$\pi\left[\frac{1}{2}a_0^2 + \sum_{k=1}^{\infty} a_k^2 + b_k^2\right] = \|f\|^2 . \tag{13}$$

Proposition 5. *(Also called the Riesz-Fischer Theorem because the completeness of L^2 is the essential ingredient in this proof.) If $\{a_n\}$, $\{b_n\}$ are sequences in ℓ^2, then*

$$\frac{1}{2}a_0 + \sum_{k=1}^{\infty} a_k \cos kx + b_k \sin kx \tag{14}$$

is the Fourier series of a function $f \in L^2$. If $s_n(x)$ is the nth partial sum of the series (13), then $s_n \longrightarrow f$ in L^2; i.e., $\|s_n - f\| \longrightarrow 0$. The equality (13) holds between the ℓ^2 norm of the coefficients and the L^2 norm of f.

Proof. We show first that if $s_n(x)$ is the nth partial sum of (13), then $\{s_n\}$ is a Cauchy sequence in L^2, and hence has a limit f. Then we show that (14) is the Fourier series of this limit f. A calculation like that of Problem 5 shows that

$$\begin{aligned}
\|s_n - s_m\|^2 &= \int \left(\sum_{k=m+1}^{n} a_k \cos kx + b_k \sin kx \right)^2 \\
&= \pi \sum_{k=m+1}^{n} (a_k^2 + b_k^2)\,,
\end{aligned}$$

all the cross product integrals being zero. Since $\{a_n\}, \{b_n\} \in \ell^2$, the identity above shows that $\{s_n\}$ is a Cauchy sequence in L^2. Since L^2 is complete, there is $f \in L^2$ so $\|s_n - f\| \longrightarrow 0$. Since $\|f_n - s_n\| \longrightarrow 0$ it follows that the inner product $(f - s_n, g) \longrightarrow 0$ for every $g \in L^2$. In particular, if $g(x) = \cos kx$, we have

$$\begin{aligned}
0 &= \lim_{n\to\infty} (f - s_n, \cos kx) \\
&= \lim_{n\to\infty} [(f, \cos kx) - (s_n, \cos kx)] \\
&= (f, \cos kx) - \pi a_k\,.
\end{aligned}$$

Therefore a_k is the kth Fourier cosine coefficient, and a similar calculation shows that b_k is the kth sine coefficient. That is, the general trigonometric series (14) formed from ℓ^2 sequences $\{a_n\}$, $\{b_n\}$, is the Fourier series of the L^2 function which is the

L^2 limit of its partial sums. We know from Bessel's inequality that

$$\|s_n\|^2 = \pi\left[\frac{1}{2}a_0^2 + \sum_{k=1}^{n}(a_k^2 + b_k^2)\right] \leq \|f\|^2 .$$

Since $\|s_n - f\| \longrightarrow 0$, it follows that $\|s_n\| \longrightarrow \|f\|$; *i.e.*,

$$\pi\left[\frac{1}{2}a_0^2 + \sum_{k=1}^{\infty} a_k^2 + b_k^2\right] = \|f\|^2 . \quad ▮$$

To sum up: each ℓ^2 sequence defines a trigonometric series whose partial sums converge in L^2 to an L^2 function f. The Fourier coefficients of this f are the given sequence in ℓ^2. The L^2 functions all determine sequences of Fourier coefficients in ℓ^2, and the above result says this mapping from L^2 to ℓ^2 is onto. The mapping is obviously linear, since integrals are linear. The mapping is also one-to-one. This is our one article of faith. There is, therefore, a one-to-one linear mapping φ on L^2 onto ℓ^2 such that $\pi\|\varphi(f)\|^2 = \|f\|^2$ for all $f \in L^2$.

INDEX